Proofs 101

Proofs 101

An Introduction to Formal Mathematics

Joseph Kirtland

CRC Press
Taylor & Francis Group
Boca Raton London New York

CRC Press is an imprint of the
Taylor & Francis Group, an **informa** business

First edition published 2021
by CRC Press
6000 Broken Sound Parkway NW, Suite 300, Boca Raton, FL 33487-2742

and by CRC Press
2 Park Square, Milton Park, Abingdon, Oxon, OX14 4RN

Library of Congress Cataloging-in-Publication Data

Names: Kirtland, Joseph (Mathematics professor), author.
Title: Proofs 101 : an introduction to formal mathematics / Joseph Kirtland.
Description: First edition. | Boca Raton : C&H/CRC Press, 2020. | Includes bibliographical references and index.
Identifiers: LCCN 2020027997 (print) | LCCN 2020027998 (ebook) | ISBN 9780367536930 (hardback) | ISBN 9780367536817 (paperback) | ISBN 9781003082927 (ebook)
Subjects: LCSH: Proof theory. | Logic, Symbolic and mathematical.
Classification: LCC QA9.54 .K57 2020 (print) | LCC QA9.54 (ebook) | DDC 511.3/6--dc23
LC record available at https://lccn.loc.gov/2020027997
LC ebook record available at https://lccn.loc.gov/2020027998

ISBN: 9780367536930 (hbk)
ISBN: 9780367536817 (pbk)
ISBN: 9781003082927 (ebk)

To my mother, Monika Hornsteiner Kirtland, and the memory of my father, Lynn Kirtland.

Contents

Preface

0.1   TO THE STUDENT

Up to this point in your mathematical career, you have completed, or nearly completed, the calculus sequence and have most likely taken a first course in linear algebra. In addition to emphasizing the beauty and power of mathematics, the main goal of these courses was to develop computational skills, improve problem-solving techniques, and introduce ideas central to your development as a mathematician. However, to continue to mathematically mature and grow, you need to move beyond being computationally proficient and learn how to create and prove the mathematical results you are using.

So what is a proof? Simply put, a proof is a valid argument that demonstrates that a mathematical statement is true. You have been writing informal proofs for years and may not have known it. For example, you are given the equation

$$x^2 - 3x - 4 = 0$$

and are asked to solve for x. The first step is to factor $x^2 - 3x - 4$. You know when multiplying two binomials $(x + a)$ and $(x + b)$ together, you get

$$(x + a)(x + b) = x^2 + (a + b)x + ab.$$

Thus you are looking for two numbers a and b such that $a + b = -3$ and $ab = -4$. A quick examination of the possibilities shows that $a = 1$ and $b = -4$. You can then write the equation

$$x^2 - 3x - 4 = 0$$

as

$$(x + 1)(x - 4) = 0.$$

At this point, you apply the result for numbers c and d that if $cd = 0$, then $c = 0$ or $d = 0$. Thus $(x + 1)(x - 4) = 0$ indicates that $x + 1 = 0$

or $x - 4 = 0$. Some quick algebra gets you $x = -1$ or $x = 4$. You have just proven the statement: "If $x^2 - 3x - 4 = 0$, then $x = -1$ or $x = 4$." In this book, a variety of proof techniques will be introduced and you will slowly build up your ability to prove mathematical statements. The proof techniques you will learn can be applied in any area of mathematics. Consequently, this book will prepare you for all of your upper-level mathematics courses.

So why do we care about proofs? Shouldn't we just accept a mathematical statement as true because it appears correct? If this was the case, then according to Aristotle (ca. 384–322 BCE), heavier objects would fall faster than lighter objects. This is definitely not what happens as shown by Galileo Galilei (1564–1642). The story goes that he dropped two spheres, each of a different mass, at the same time from the top of the Leaning Tower of Pisa and that they simultaneously hit the ground. While this is probably not true, he did observe the rate at which two spheres of the same size, one of lead and one of cork, rolled down an inclined ramp [Gal]. In addition, based on his observations, Galileo determined that the distance traveled by an object in free fall is proportional to the square of the time traveled and derived a formula for free-fall motion. If you are interested in the historic development of this topic, I suggest [Dra89].

The work done by Galileo on free-fall motion highlights another important aspect of mathematics. Mathematics is in some sense a lab science where results do not simply spring out of thin air. They come from knowing what has been done in the past, by completing numerous and varied examples, and recognizing observable patterns. A large amount of experimentation goes into a mathematical conclusion. Only after you believe you have enough evidence do you then state your result and write a proof to justify it.

So who are proofs for? They are for you and the mathematical world. You write a proof to convince yourself that the result you have developed or are using is true. In addition, writing a proof of a result forces you to digest every aspect of the result you are proving, helping you understand the motivation for, the mathematics behind, and the consequences of the result. The thought process involved helps you gain a deeper appreciation of the result and better understanding of the mathematics involved. Simply put, you are learning when you are writing a proof.

In addition, you write a proof for the mathematical community. You want to communicate to the world that the result you have is true and can be used. You want others to read your proof and understand it.

This means that your proof needs to be grammatically and rhetorically sound, mathematically correct, and well organized.

So how does one write a proof? Well, continue reading as this is the goal of the book. However, simply put, a proof is a narrative that clearly explains to the reader why a result is true. The reader should be able to follow every step. With this in mind, here are a few global guidelines. We will get into the specifics in the chapters to follow.

- As mentioned above, proofs are to be written in complete, grammatically correct sentences. The rules of syntax and punctuation still apply.

- You can write mathematical symbols, expressions, and equations in sentences, but avoid starting a sentence with them. For example, instead of writing, "$x^2 - 3x - 4 = 0$ has solutions $x = -1$ and $x = 4$," write, "The equation $x^2 - 3x - 4 = 0$ has solutions $x = -1$ and $x = 4$."

- Never use a symbol to replace a word in a sentence that is not part of a mathematical expression. For example, don't write, "Given $=$ integers a and b, we know...," instead write, "Given that integers a and b are equal, we have..," or "Given that $a = b$, for integers a and b, we have..."

- You must define all variables when you first use them. For example, if you want to introduce an expression concerning sum of two integers, write, "Consider the sum $a + b$, for integers a and b" or "For integers a and b, consider the sum $a + b$."

Just as with anything else, to learn a new skill you need to get your hands dirty. Mathematics is not a spectator sport. Practice and do not be afraid to stumble. Some of the best learning experiences come from not getting it right. As Robert F. Kennedy famously said,

Only those who dare to fail greatly can ever achieve greatly.

Finally, let me say that in addition to teaching you how to write proofs, this text will also introduce a number of concepts at the core of the upper-level mathematics classes you will take. Pay attention to the definitions presented, the topics discussed, and the results proven as you will see a number of them in the future. You will also be introduced to new notation and new terminology that will be used in other courses. Upon working through this text, you will be prepared to tackle upper-level mathematics courses.

0.2 TO THE PROFESSOR

This text introduces students to the proof-writing techniques needed for completing upper-level undergraduate mathematics courses. Given that the content for most of the proofs they will write comes from number theory, set theory, and analysis of functions, the students will also learn concepts at the core of their junior- and senior-level mathematics classes. Furthermore, this text will also introduce terminology, notation, and some of the results that they will see and use as they continue their study of mathematics.

While this book has a number of features in common with other textbooks aimed at the same topic, there are a number of distinctive features:

- The text begins with a short chapter on logic that introduces all of the material needed for developing proof-writing techniques.

- Proof techniques are introduced in the second chapter of the text. The goal here is to get the students writing proofs early and often.

- The chapter on sets comes after proof techniques are introduced so students can apply the their proof-writing skills to justifying results from set theory as they learn them.

- All important results are explicitly stated in the text and do not appear for the first time in the exercises. While the students will still prove many or parts of these results, they appear in the body of the text for easy reference.

- Given the content of the book, it is easy to cover all of the material in a single 16-week semester.

- At the end of most sections, there is a wealth of exercises ranging in difficulty from very accessible to more formidable.

- A number of exercises are crafted to prepare the students for concepts learned in subsequent sections of the text. For example, students complete a number of exercises concerning the range and preimage of a function that are designed to prepare them for the concept of a surjective function.

The seven chapters in this text cover the topics listed here. They are written with the intent that they be covered in a linear order.

Chapter 1 begins the text with a brief introduction to symbolic and predicate logic. It introduces the students to statements, predicates, quantifiers, and logical connectives. Truth tables are introduced and used to show when statements are logically equivalent. A number of important logical equivalences needed in subsequent chapters are given.

The students start writing proofs in Chapter 2. It begins with an introduction to the axiomatic nature of mathematics with the importance of definitions being emphasized. In addition, a variety of number theoretic concepts, such as parity, divisibility, prime numbers, and modulo n, are developed. Rational and irrational numbers are also introduced. The chapter focuses on the proof methods of direct proof, proof by contrapositive, proof by cases, and proof by contradiction.

Chapter 3 presents a naive approach to set theory. Once the concept of a set is agreed upon, the concepts of set equality, subset, union, intersection, difference, and complement are introduced. The students begin applying their newly learned proof techniques in this chapter as they prove a number of set theoretic results. This includes an examination of indexed sets. The chapter ends with a brief discussion of Russell's Paradox and its impact.

The students are introduced to proof by mathematical induction in Chapter 4. This chapter includes a section on strong induction, using it as an opportunity to introduce and discuss the Fibonacci sequence.

Relations are examined in Chapter 5. After an introduction to relations, the chapter focuses on the three properties of reflexive, symmetric and transitive relations, leading to a study of equivalence relations and equivalence classes.

Using the language of relations, functions are discussed in Chapter 6. This immediately leads to an investigation of injective, surjective, and bijective functions. Composition of functions and the inverse of a function are studied, with the chapter ending with the result that a function is bijective if and only if it has an inverse that is also bijective.

The text ends with Chapter 7 on the cardinality of sets. After defining and investigating what it means for two sets to have the same cardinality, fundamental results concerning finite, infinite, countable, and uncountable sets are established. The chapter ends with a section on comparing cardinalities, with the main focus being Cantor's result concerning power sets, Cantor's Paradox, and the Continuum Hypothesis.

The transition from lower-level to upper-level mathematics classes is an important step in the development of any mathematician. This text will help students to smoothly make that passage.

Acknowledgments

This book would not have been possible without the input from the numerous Marist College students I have taught over the years in our "Introduction to Mathematical Reasoning" course. The order of topics, the approach used to teach proof writing, and the diverse exercises found in this text were influenced by student comments, contributions, and constructive criticism. In this way, each class and student I have taught has helped to hone this text. They have taught me so much.

I would specifically like to thank James Helmreich, Homer Bechtell, and Marist student, Madison Russell, for their eagle eyes, insights, and extremely helpful comments that improved this book.

My appreciation also goes to Callum Fraser, editor of mathematics books, and Mansi Kabra, editorial assistant, at CRC Press/Taylor & Francis Group for all of their efforts.

Symbol Description

Logic

1.1 INTRODUCTION

Logic is at the foundation of everything we do in mathematics. It provides a formal language that is used to state and process results and a system of deductive reasoning used to justify or prove them. With the goal of this book in mind, the tools learned from logic are essential for crafting, writing, and analyzing proofs.

This chapter introduces a few of the basic concepts in logic. While this is not an in-depth study, it does provide all of the tools we will need in subsequent chapters. However, once you finish reading this chapter and are interested in learning more about logic, I suggest [O'L16] and [GU89].

1.2 STATEMENTS AND LOGICAL CONNECTIVES

Statements are the basic elements of logic.

Definition 1.2.1. *A **statement** is a declarative sentence that is either true or false, but not both.*

To simplify the calculus of logic, the letters $P, Q, R, S, \ldots$ are used to denote statements. The following sentences are all statements.

$P:$ $7 + 4 = 12$.

$Q:$ The Empire State Building is in New York City.

$R:$ The 2^{100} digit of π is 3.

The statement P is false, Q is true, and R, while we may not know the 2^{100} digit of π, is definitely either true or false.

Sentences such as, "Are you tired," "$x + 3 = 11$," and "The Empire State Building is tall," are not statements. The first is not a statement as it is a question. The second is not as the value of x is unknown. The last sentence is trickier. Philosophers might say that if a universal definition of the word "tall" can be agreed to, then it would be a statement. However, we will avoid these ambiguities.

Individual statements can be combined, using five **logical connectives**, to create new ones called **compound statements**. When examining compound statements consisting of logical connectives and individual statements, the letters $P, Q, R, S, \ldots$ used to denote individual statements are viewed as variables that take on the value T for true or F for false.

Definition 1.2.2. *Given a statement P, the **negation** of P is the statement "not P" denoted by $\sim P$.*

For example, if P is the statement, "$2^3 + 1 < 10$," then $\sim P$ is the statement, "$2^3 + 1 \geq 10$." And if Q is the statement, "Golf balls are cube-shaped," then $\sim Q$ is the statement, "Golf balls are not cube-shaped." When a statement is true, as P above is, then the negation of the statement is false. When a statement is false, as Q above is, then the negation of the statement is true. This observation can be represented in a **truth table**. A truth table gives the truth values of a compound statement based on the truth value combinations of the individual statements that are in it. Given a statement P, the truth table for $\sim P$ is given here.

P	$\sim P$
T	F
F	T

Now on to the next two logical connectives.

Definition 1.2.3. *Let P and Q be statements.*

(i) *The **conjunction** of P and Q is the statement "P and Q" denoted by $P \wedge Q$.*

(ii) *The **disjunction** of P and Q is the statement "P or Q" denoted by $P \vee Q$.*

The statement $P \wedge Q$ is only true when both P and Q are true. For example, if P is the statement, "Cindy is wearing boots" and Q is the statement, "Cindy is wearing a hat," the statement $P \wedge Q$: "Cindy is wearing boots and a hat" is true only when P and Q are both true. If Cindy is not wearing boots (P is false) or not wearing a hat (Q is false), then the statement $P \wedge Q$ is false, as it is not true that Cindy is wearing boots and a hat. This results in the following truth table for $P \wedge Q$.

P	Q	$P \wedge Q$
T	T	T
T	F	F
F	T	F
F	F	F

We need to spend a bit more time on the disjunction. It is not to be understood in the colloquial sense of the word "or." For example, if you say, "This summer I will work as a lifeguard on Cayuga Lake or work as an intern in New York City," you are using the exclusive form of the word "or." In other words, you will do one or the other, but not both. In logic, the word "or" is used inclusively. For example the statement, "It will snow today or tomorrow" implies that it could snow both days. The statement is true if it snows both today and tomorrow. Given statements P and Q, this results in the following truth table for $P \vee Q$.

P	Q	$P \vee Q$
T	T	T
T	F	T
F	T	T
F	F	F

The fourth logical connective is now defined.

Definition 1.2.4. *Given the statements P and Q, the **conditional** or **implication** is the statement "if P, then Q" or "P implies Q" denoted $P \to Q$.*

To determine the truth values of the conditional, it helps to view it as a promise. For example, let P be the statement, "I lost the bet," and let Q be the statement, "I will shave my head." Thus the statement $P \to Q$ is, "If I lose the bet, then I will shave my head." This statement will be true if I live up to my promise and false if I do not. If I do lose the bet (P is true) and then I go ahead and shave my head (Q is true), then clearly I have kept my promise and $P \to Q$ is true. However, if I lose the bet (P is true) and I fail to shave my head (Q is false), then I have not kept my promise and $P \to Q$ is false. Now suppose that I did not lose the bet (P is false). Then according to my promise, I do not need to do anything. Regardless of whether I shave my head or not, I have kept my promise, which is to take action if I lose the bet. Thus in this situation, the statement $P \to Q$ will be true. Thus for statements P and Q, the truth table for $P \to Q$ is as follows:

P	Q	$P \to Q$
T	T	T
T	F	F
F	T	T
F	F	T

When the statement P is false, the conditional statement $P \to Q$ is always true. When this occurs, we say that $P \to Q$ is **vacuously true**. In the statement "if P, then Q," the statement P is referred to as the **hypothesis** or **antecedent** and Q is referred to as the **conclusion** or **consequent**. In addition, the following list provides a number of different ways to express $P \to Q$:

(i) If P, then Q.   (iii) P only if Q.   (v) P is sufficient for Q.

(ii) P implies Q.   (iv) Q if P.   (vi) Q is necessary for P.

We have now arrived at the last logical connective.

Definition 1.2.5. *Given the statements P and Q, the* **biconditional** *is the statement "P if and only if Q" denoted $P \leftrightarrow Q$.*

The statement $P \leftrightarrow Q$ implies a logical equality between the two statements P and Q. It means that $P \to Q$ and $Q \to P$. Given this sense of equality, $P \leftrightarrow Q$ is only true when both P and Q have the same truth value, otherwise it is false. Given statements P and Q, the truth table for $P \leftrightarrow Q$ is as follows.

P	Q	$P \leftrightarrow Q$
T	T	T
T	F	F
F	T	F
F	F	T

The following list documents identical ways to express $P \leftrightarrow Q$:
(i) P if and only if Q.
(ii) P is equivalent to Q.
(iii) P is necessary and sufficient for Q.

Exercises 1.2

(1) Determine which of the following sentences are statements. If the sentence is a statement, indicate whether it is true or false.

(a) The Boston Red Sox play at Fenway Park.

(b) The best color for a basketball is orange.

(c) How old are you?

(d) February 3, 2030 will be a Monday.

(e) The numbers 2, 4, and 6 are the only even integers.

(f) $x^2 \geq 0$.

(2) Determine which of the following statements are true and which are false.

(a) Soccer balls are not spherical.

(b) The sun is yellow and the sky is blue.

(c) If the sun is blue, then the sky is yellow.

(d) $(e^2 < 4) \wedge (3^2 > 7)$

(e) $(e^2 < 4) \wedge (3^2 < 7)$

(f) $(\sqrt{3} > 1) \vee \left(\sin\left(\frac{\pi}{8}\right) < -2\right)$

(g) $(\sqrt{3} < 1) \vee \left(\sin\left(\frac{\pi}{8}\right) > 2\right)$

(h) $(\sqrt{9} = 3) \rightarrow (e^2 < 4)$

(i) $(\sqrt{9} = -3) \rightarrow (e^2 < 4)$

(j) $(\sqrt{3} < 1) \leftrightarrow (3^2 > 1)$

(k) $(\sqrt{3} < 1) \leftrightarrow (3^2 < 1)$

1.3 LOGICAL EQUIVALENCE

Section 1.2 has embarked us down the path of symbolic logic. An entire semester alone could be spent on the subject. In this section we will introduce the tools from symbolic logic needed in subsequent chapters.

As introduced in Section 1.2, individual statements can be combined using the five logical connectives to create compound statements. Viewed as operations, there are rules for the order in which the logical connectives are applied when creating a compound statement with more than one logical connective. The order of operations is as follows:

(1) negation ($\sim$)

(2) conjunction and disjunction ($\wedge$ and $\vee$)

(3) implication ($\rightarrow$)

(4) biconditional ($\leftrightarrow$)

Parentheses may be used to clear up any ambiguity. For example, the compound statement

$$\sim P \vee Q \rightarrow \sim Q \wedge R$$

means

$$((\sim P) \vee Q) \rightarrow ((\sim Q) \wedge R).$$

To start, recall the conditional statement $P \rightarrow Q$ for statements P and Q. Associated with it are the following three compound statements.

- The **converse** of $P \rightarrow Q$ is the statement $Q \rightarrow P$.

- The **contrapositive** of $P \rightarrow Q$ is the statement $\sim Q \rightarrow \sim P$.

- The **inverse** of $P \rightarrow Q$ is the statement $\sim P \rightarrow \sim Q$.

Truth tables are used to determine the truth value of a compound statement for each of the truth value combinations of the individual statements that it contains. The compound statement $(\sim P \vee Q) \rightarrow P \wedge Q$ has the following truth table, indicating the truth value of the statement for all of the different truth value combinations of P and Q.

P	Q	$\sim P$	$\sim P \vee Q$	$P \wedge Q$	$(\sim P \vee Q) \rightarrow P \wedge Q$
T	T	F	T	T	T
T	F	F	F	F	T
F	T	T	T	F	F
F	F	T	T	F	F

For example, when P is true and Q is false, going to the end of the second row of the truth table (highlighted in gray) we see that the compound statement will be true.

There are specific compound statements that are of interest to mathematicians. Consider the compound statement $P \vee \sim P$ and its truth table.

P	$\sim P$	$P \vee \sim P$
T	F	T
F	T	T

This compound statement is always true. A compound statement that is true for all possible truth value combinations of the individual statements that it contains is called a **tautology**. Tautologies will be important as we learn to write proofs. Given the statements P and Q, the statement $(P \wedge (P \rightarrow Q)) \rightarrow Q$ is also a tautology as demonstrated by its truth table.

P	Q	$P \to Q$	$P \wedge (P \to Q)$	$(P \wedge (P \to Q)) \to Q$
T	T	T	T	T
T	F	F	F	T
F	T	T	F	T
F	F	T	F	T

Here is another example. The truth table below demonstrates that given the statements P, Q, and R, the compound statement $(P \to Q) \wedge (Q \to R) \to (P \to R)$ is a tautology with S_1 denoting $(P \to Q) \wedge (Q \to R)$ and S_2 denoting $P \to R$

P	Q	R	$P \to Q$	$Q \to R$	S_1	$P \to R$	$S_1 \to S_2$
T	T	T	T	T	T	T	T
T	T	F	T	F	F	F	T
T	F	T	F	T	F	T	T
T	F	F	F	T	F	F	T
F	T	T	T	T	T	T	T
F	T	F	T	F	F	T	T
F	F	T	T	T	T	T	T
F	F	F	T	T	T	T	T

Theorem 1.3.1 lists three fundamental tautologies. The first statement was shown earlier to be a tautology. Showing the second and third statements are tautologies is an exercise.

Theorem 1.3.1. *For statements P and Q, the following compound statements are tautologies.*

   (i) $P \vee \sim P$ *Excluded Middle*

   (ii) $P \to P \vee Q$ *Addition*

   (iii) $P \wedge Q \to P$ *Simplification*

Also of interest are compound statements that are always false. These are called **contradictions**. For example, the truth table for the compound statement $P \wedge \sim Q \leftrightarrow (P \to Q)$ shows it is a contradiction.

P	Q	$\sim Q$	$P \wedge \sim Q$	$P \to Q$	$P \wedge \sim Q \leftrightarrow (P \to Q)$
T	T	F	F	T	F
T	F	T	T	F	F
F	T	F	F	T	F
F	F	T	F	T	F

Note that when statement C is a contradiction, then $\sim C$ is a tautology. As we shall see, contradictions also play a role in mathematics. It should also be noted that a compound statement that is neither a tautology nor a contradiction is called a **contingency**.

When two statements S_1 and S_2 have identical truth values for the same combination of truth values of the individual statements that are contained in each, they are called **logically equivalent**, denoted by $S_1 \equiv S_2$. For example, the compound statements $P \to Q$ and $\sim P \vee Q$, as shown in the truth tables below, are logically equivalent.

P	Q	$P \to Q$
T	T	T
T	F	F
F	T	T
F	F	T

P	Q	$\sim P$	$\sim P \vee Q$
T	T	F	T
T	F	F	F
F	T	T	T
F	F	T	T

In addition, given that $S_1 \leftrightarrow S_2$ will only be true when the two statements S_1 and S_2 are both true or when they are both false, they also can be shown to be logically equivalent if the compound statement $S_1 \leftrightarrow S_2$ is a tautology. The fact that $P \to Q \equiv \sim P \vee Q$ can also be demonstrated with the following truth table.

P	Q	$P \to Q$	$\sim P$	$\sim P \vee Q$	$P \to Q \leftrightarrow \sim P \vee Q$
T	T	T	F	T	T
T	F	F	F	F	T
F	T	T	T	T	T
F	F	T	T	T	T

A list of regularly used logical equivalences is given in Theorem 1.3.2.

Theorem 1.3.2. *Let P, Q, and R be statements.*

(i)	$\sim (\sim P) \equiv P$	*Double Negation*
(ii)	$P \wedge Q \equiv Q \wedge P$	*Commutative Law*
(iii)	$P \vee Q \equiv Q \vee P$	*Commutative Law*
(iv)	$(P \wedge Q) \wedge R \equiv P \wedge (Q \wedge R)$	*Associative Law*
(v)	$(P \vee Q) \vee R \equiv P \vee (Q \vee R)$	*Associative Law*
(vi)	$P \wedge (Q \vee R) \equiv (P \wedge Q) \vee (P \wedge R)$	*Distributive Law*

(vii) $P \vee (Q \wedge R) \equiv (P \vee Q) \wedge (P \vee R)$ *Distributive Law*

(viii) $P \to Q \equiv \sim Q \to \sim P$ *Contrapositive*

(ix) $\sim (P \wedge Q) \equiv \sim P \vee \sim Q$ *De Morgan's Law*

(x) $\sim (P \vee Q) \equiv \sim P \wedge \sim Q$ *De Morgan's Law*

(xi) $P \to Q \equiv \sim P \vee Q$

(xii) $P \leftrightarrow Q \equiv (P \to Q) \wedge (Q \to P)$

(xiii) $\sim (P \to Q) \equiv P \wedge \sim Q$

The truth table created earlier for $P \to Q \leftrightarrow \sim P \vee Q$ shows that statement (xi) from Theorem 1.3.2 is true. Since showing two statements are logically equivalent simply requires a truth table, only statement (vi) from Theorem 1.3.2 will be proven. A number of the others are exercises.

Proof of Theorem 1.3.2 (vi). Given that the truth table, constructed below, for the statement $P \wedge (Q \vee R) \leftrightarrow (P \wedge Q) \vee (P \wedge R)$, with S_1 denoting $P \wedge (Q \vee R)$ and S_2 denoting $(P \wedge Q) \vee (P \wedge R)$, is a tautology, the statements $P \wedge (Q \vee R)$ and $(P \wedge Q) \vee (P \wedge R)$ are logically equivalent.

P	Q	R	$Q \vee R$	S_1	$P \wedge Q$	$P \wedge R$	S_2	$S_1 \leftrightarrow S_2$
T	T	T	T	T	T	T	T	T
T	T	F	T	T	T	F	T	T
T	F	T	T	T	F	T	T	T
T	F	F	F	F	F	F	F	T
F	T	T	T	F	F	F	F	T
F	T	F	T	F	F	F	F	T
F	F	T	T	F	F	F	F	T
F	F	F	F	F	F	F	F	T

□

Exercises 1.3

(1) Create a truth table for each of the following statements. Indicate if the statement is a tautology, contradiction, or a contingency.

(a) $P \wedge \sim P$ (c) $(P \wedge (P \vee Q \to R)) \to R$

(b) $(P \vee Q) \to \sim (P \wedge Q)$

(2) Let P and Q be statements. Show that the following compound statements are tautologies.

(a) $\sim Q \wedge (P \rightarrow Q) \rightarrow \sim P$

(b) $P \wedge Q \rightarrow P \vee Q$

(c) $(P \leftrightarrow Q) \leftrightarrow (P \rightarrow Q) \wedge (\sim P \rightarrow \sim Q)$

(3) Verify that the following statements from Theorem 1.3.1 are tautologies.

(a) $P \rightarrow P \vee Q$ (b) $P \wedge Q \rightarrow P$

(4) Verify the following logical equivalences from Theorem 1.3.2.

(a) $\sim (\sim P) \equiv P$ (d) $P \rightarrow Q \equiv \sim Q \rightarrow \sim P$

(b) $P \wedge Q \equiv Q \wedge P$ (e) $\sim (P \wedge Q) \equiv \sim P \vee \sim Q$

(c) $(P \wedge Q) \wedge R \equiv P \wedge (Q \wedge R)$ (f) $P \rightarrow Q \equiv \sim P \vee Q$

1.4 PREDICATES AND QUANTIFIERS

Recall the sentence "$x + 3 = 11$" introduced at the beginning of Section 1.2. This declarative sentence is not a statement, but if x is assigned a value, the sentence becomes a statement that is either true (for $x = 8$) or false (for $x = 9$). In logic, a **predicate** is a declarative sentence that contains a finite number of variables such that the sentence becomes a statement when the variables are assigned specific values.

Predicates are denoted using capital letters, typically $P, Q, R, S, \ldots$, followed by the variables contained in the predicate in parentheses. For example, let $P(x)$ denote the predicate "$x + 3 = 11$." Then $P(8)$ would be the statement "$8 + 3 = 11$," which is true. For the predicate $Q(x, y) : x^2 + y^2 = 1$, the statement $Q(1, 1)$, which is $1^2 + 1^2 = 1$, would be false.

Consider the predicates $P(x) : x^2 + 1 \geq 1$ and $Q(x) : x + 1 < -2$. The sentence, "For all real numbers x, $x^2 + 1 \geq 0$" is a statement as $P(x)$ is true for any real number x. The sentence, "For all real numbers x, $x + 1 < -2$" is also a statement, but is false as $Q(5)$ is false. However, the sentence, "There exists a real number x, such that $x + 1 < -2$" is a true statement since $Q(-4)$ is true. This motivates the following definition.

Definition 1.4.1. Let $P(x)$ be a predicate with variable x, and let U be a collection of objects.

(i) *The **universal quantifier** is the phrase, "**For all x in U**" or "**For every x in U**", denoted $\forall x \in U$, with the **universally quantified predicate** being the statement $\forall x \in U, P(x)$.*

(ii) *The **existential quantifier** is the phrase, "**There exists an x in U**" or "**There is an x in U**", denoted $\exists x \in U$, with the **existentially quantified predicate** being the statement $\exists x \in U, P(x)$.*

The statement $\forall x \in U, P(x)$ is only true if $P(x)$ is true for every value of x in U. If there is a value of x in U, say $x = a$, such that $P(a)$ is false, then the statement is false. The statement $\exists x \in U, P(x)$ is true when there is at least one value of x in U, say $x = b$, such that $P(b)$ is true. If $P(x)$ is false for every value of x in U, then the statement is false. The collection of objects U is sometimes referred to as the **universe**, as it contains all of the objects that are allowed to be considered. The following collections of numbers will be regularly used as the universe.

- The natural numbers $\mathbb{N} = \{1, 2, 3, 4, 5, \ldots\}$

- The integers $\mathbb{Z} = \{\ldots, -3, -2, -1, 0, 1, 2, 3, \ldots\}$

- The rational numbers $\mathbb{Q} = \{\frac{a}{b} : a \text{ and } b \text{ are integers with } b \neq 0\}$

- The real numbers $\mathbb{R}$

Both the quantifier and the universe used are important for determining the truth value of a quantified statement. For example, the statement $\forall x \in \mathbb{R}, x^2 + 1 \geq 2$ is false (let $x = 0$), while the statement $\forall x \in \mathbb{N}, x^2 + 1 \geq 2$ is true. When the first statement is changed to $\exists x \in \mathbb{R}, x^2 + 1 \geq 2$, it is then true. However, note that in general changing the quantifier does not change the truth value of a quantified predicate. The statement $\forall x \in \mathbb{R}, x^2 + 1 \leq -2$ is false with the statement $\exists x \in \mathbb{R}, x^2 + 1 \leq -2$ also being false.

A predicate can be quantified by two or more quantifiers over different variables. The truth value of such a statement can be sensitive to the order of the quantifiers. For example, consider the statement

$$\forall x \in \mathbb{Z}, \exists y \in \mathbb{Z}, x + 1 = y.$$

This statement is true. For any possible integer x, the expression $x + 1$ is also an integer. Thus for every integer x, there exists an integer y such that $x + 1 = y$. Now consider the statement

$$\exists x \in \mathbb{Z}, \forall y \in \mathbb{Z}, x + 1 = y.$$

For this statement to be true, there must exist a fixed integer x (there could be more than one) such that $x + 1$ takes on the value of every integer y. Since it is not possible for a fixed integer to equal all of the other integers, this statement is false.

When two (or more) of the same quantifier and universe are used, this statement can by symbolically shortened. The statement

$$\forall x, y \in U, P(x, y) \text{ denotes } \forall x \in U, \forall y \in U, P(x, y)$$

and

$$\exists x, y \in U, P(x, y) \text{ denotes } \exists x \in U, \exists y \in U, P(x, y).$$

The five logical connectives are also used in quantified statements. For example, consider the statement

$$\forall x \in \mathbb{R}, (x^2 > 1) \to (x > 1).$$

This statement is false as shown by the fact that $x = -2$ makes the statement $x^2 > 1$ true, but the statement $x > 1$ false. Another example is

$$\exists r, s \in \mathbb{Z}, (rs > 0) \wedge \sim (r + s > 0).$$

Let $r = -2$ and $s = -3$. Clearly, the statement $rs > 0$ is true. Given that the statement $r + s > 0$ is false, the statement $\sim (r + s > 0)$ is true. Therefore, $(rs > 0) \wedge \sim (r + s > 0)$ is true and the statement is true.

Exercises 1.4

(1) Determine which of the following statements are true and which are false. Provide a brief explanation to support your answer.

(a) $\forall x \in \mathbb{R}, e^x + 1 > 1$.

(b) $\forall n \in \mathbb{N}, \frac{1}{n} < 1$.

(c) $\forall x \in \mathbb{R}, x^2 - 4x + 7 > 0$.

(d) $\exists z \in \mathbb{Z}, z^2 = 3$.

(e) $\exists r \in \mathbb{Q}, 5r > 1$.

(f) $\forall x, y \in \mathbb{R}, x^2 + y^2 > 1$.

(g) $\forall z \in \mathbb{Z}, \exists n \in \mathbb{N}, n < z$.

(h) $\forall n \in \mathbb{N}, \exists r \in \mathbb{R}, \frac{1}{n} < r$.

(i) $\exists z \in \mathbb{Z}, \forall n \in \mathbb{N}, z < n$.

(j) $\exists r, s \in \mathbb{Q}, r - s > 0$.

(k) $\forall x \in \mathbb{R}, (x^2 < 1) \to (x > -1) \vee (x < 1)$.

(l) $\exists n \in \mathbb{N}, (\frac{1}{n} \geq 1) \wedge (n^2 = 4)$.

(m) $\forall x, y \in \mathbb{R}, (x^2 > y^2) \to (x > y)$.

(n) $\forall x \in \mathbb{R}, \exists n \in \mathbb{N}, (x < 0) \vee (x > 0) \to x^n > 0$.

1.5 NEGATION

Given a statement S, the **negation** of S is the statement $\sim S$. There are rules than can be applied for negating certain regularly used compound statements in order to rewrite them in a logically equivalent, and perhaps more usable, form. From Theorem 1.3.2, we know that

- $\sim (\sim P) \equiv P$,

- $\sim (P \wedge Q) \equiv \sim P \vee \sim Q$,

- $\sim (P \vee Q) \equiv \sim P \wedge \sim Q$, and

- $\sim (P \rightarrow Q) \equiv P \wedge \sim Q$.

For example, the negation of the statement, "The integer 4 is odd and a perfect square" is the statement, "The integer 4 is not odd or not a perfect square." The negation can be more simply written as, "The integer 4 is even or not a perfect square."

But what about quantified statements? To see how negating a quantified statement works, let's examine the statement $\forall x \in \mathbb{R}, x^2 > 1$ or, "For all real numbers x, $x^2 > 1$," which is false. Negating this statement means that it is not the case that all real numbers x satisfy $x^2 > 1$ or that there exists a real number a that does not satisfy $a^2 > 1$. The negation is true as demonstrated by $a = \frac{1}{2}$.

Now consider the statement $\exists x \in \mathbb{R}, x^2 > 1$ or, "There exists a real number x, such that $x^2 > 1$," which is true. Negating this statement means that it is not the case that there is a real number x satisfying $x^2 > 1$ or that for every real number x, we have $x^2 \leq 1$, which is false (let $x = 2$). This leads to the following result.

Theorem 1.5.1. *Let $P(x)$ be a predicate with variable x, and let U be the universe.*

(i) $\sim (\forall x \in U, P(x)) \equiv \exists x \in U, \sim P(x)$.

(ii) $\sim (\exists x \in U, P(x)) \equiv \forall x \in U, \sim P(x)$.

Example 1.5.2. *The negation of the following statements will be determined.*

(i) *For all real numbers x, $e^x \leq 1$ or $x > 0$.*

In symbolic form, this is written as $\forall x \in \mathbb{R}, e^x \leq 1 \vee x > 0$. Thus the negation of the statement is determined as follows.

$$\sim (\forall x \in \mathbb{R}, e^x \leq 1 \vee x > 0) \equiv \exists x \in \mathbb{R}, \sim (e^x \leq 1 \vee x > 0)$$
$$\equiv \exists x \in \mathbb{R}, \sim (e^x \leq 1) \wedge \sim (x > 0)$$
$$\equiv \exists x \in \mathbb{R}, e^x > 1 \wedge x \leq 0$$

The negation is the statement, "There exists a real number x such that $e^x > 1$ and $x \leq 0$."

(ii) There exists an integer n such that for all real numbers y, if $y^n > 1$ then $y > 1$.

In symbolic form, this is written $\exists n \in \mathbb{Z}, \forall y \in \mathbb{R}, y^n > 1 \rightarrow y > 1$. Thus the negation of the statement is determined as follows.

$$\sim (\exists n \in \mathbb{Z}, \forall y \in \mathbb{R}, y^n > 1 \rightarrow y > 1)$$
$$\equiv \forall n \in \mathbb{Z}, \sim (\forall y \in \mathbb{R}, y^n > 1 \rightarrow y > 1)$$
$$\equiv \forall n \in \mathbb{Z}, \exists y \in \mathbb{R}, \sim (y^n > 1 \rightarrow y > 1)$$
$$\equiv \forall n \in \mathbb{Z}, \exists y \in \mathbb{R}, y^n > 1 \wedge \sim (y > 1)$$
$$\equiv \forall n \in \mathbb{Z}, \exists y \in \mathbb{R}, y^n > 1 \wedge y \leq 1$$

The negation is the statement, "For all integers n, there exists a real number y such that $y^n > 1$ and $y \leq 1$."

Exercises 1.5

(1) Determine the negation of the following statements.

(a) The integer 32 is even and a power of 2.

(b) For all integers n, if n is even, then n is a power of 2.

(c) There exists a real number x, such that $x^2 > 1$ and $x < 0$.

(d) There exists integers n and m, such that $2^n + 3^m$ is a prime number.

(e) For all integers n and m, if $n < 0$ and $m < 0$, then $nm > 0$.

(f) For every real number x, there exists a natural number n, such that if $x > 0$, then $n \cdot |x| > 10$.

CHAPTER 2

Proof Techniques

2.1 INTRODUCTION

A proof shows that a mathematical statement is true. Documenting mathematical observations has a long and rich history.

- A bone found in Africa, dating to 9000–6500 BCE, contains notches arranged in groups, perhaps indicating arithmetic observations.

- It is theorized that integers written on Babylonian cuneiform tablets were arranged according to mathematical properties.

- Computational exercises found on the Rhind papyrus written by the Egyptians indicate that mathematics had established properties to be learned and practiced.

- Greek mathematicians embraced the idea that all statements needed to be formally justified. Most notable is the work *Elements*, a series of 13 books written by Euclid (see [Hea02]), where he rigorously proved results in geometry and other areas.

Before we start writing proofs, the framework of modern mathematics needs to be established. However, if you are interested in the history and evolution of mathematics, I suggest [GG98] and [FG87].

2.2 AXIOMATIC AND RIGOROUS NATURE OF MATHEMATICS

To demonstrate the modern axiomatic nature of mathematics, we will reflect upon the work you did in geometry. In general, any investigation begins by settling on the undefined terms. These are the words or phrases whose meaning everyone agrees to without explanation. In geometry, the terms **point**, **line**, **on** (in the sense that a point is on a line), and **plane** are accepted as undefined terms.

The next step is to identify the properties, called **axioms** or **postulates**, that these undefined terms satisfy. They are accepted as evident or obviously true (accepted without justification) and can involve the relation between the undefined terms or a way to manipulate them. In geometry, the following four axioms might be used as foundational facts concerning points, lines, planes and the concept of on.

- *Axiom 1*: Every two distinct points determine a unique line.

- *Axiom 2*: Every line contains at least two distinct points.

- *Axiom 3*: There exist at least three points in a plane.

- *Axiom 4*: Not all points in a plane are on the same line.

Once the undefined terms and axiom or postulates have been established, the objects that will be studied need to be defined. In geometry, this might mean defining what is meant by a triangle, a rectangle, or parallel lines. For example, consider the following definition.

Definition 2.2.1. *Two lines are **parallel** if there does not exist a point on both of them.*

Typically in mathematics, the term being defined is written in bold or italics to emphasize to the reader the specific item being defined. It can't be stated strongly enough, but all of mathematics flows through definitions. After all this, results are stated and proven. For example, the following result might be proven concerning lines.

Result 2.2.2. *Given two lines in a plane, they are parallel or intersect at exactly one point.*

Results or statements of fact must be justified or shown to be true. There are different types of these statements of fact. A **proposition** is a result that contains a smaller idea or a not very weighty result. A **lemma** is similar, but is often used in the proof a theorem that follows it. A **theorem** is typically an important result concerning a big idea. A **corollary** is a result that directly follows from a previously proven theorem.

In the coming chapters, we will learn techniques used to prove propositions, lemmas, theorems, and corollaries.

2.3 FOUNDATIONS

While the proof techniques we will learn can be used in all mathematical disciplines, this chapter will concentrate on proving results from number theory, an area of mathematics focused on the study of the integers. To start, we will accept the term **integer** as undefined. In reality, we already did this in Chapter 1. Recall that an integer is any number from the collection $\mathbb{Z} = \{\ldots, -3, -2, -1, 0, 1, 2, 3, \ldots\}$. We will also accept that the operations of addition, subtraction, and multiplication are defined on the integers. More specifically, we will assume the following axiom.

Axiom 2.3.1. *Given integers a and b, then $a+b, a-b$, and ab are also integers.*

For the algebra needed to complete our investigation of the integers, we will also need the following axiom.

Axiom 2.3.2. *Let a, b, and c be integers. Then*

(i) $a + b = b + a$,

(ii) $a + (b + c) = (a + b) + c$,

(iii) $ab = ba$,

(iv) $a(bc) = (ab)c$,

(v) $a(b + c) = ab + ac$,

(vi) $(a + b)c = ac + bc$,

(vii) $a + 0 = a$ and $1 \cdot a = a$, and

(viii) $ac = bc$ implies $a = b$ for $c \neq 0$.

We will regularly use these two axioms, but will not specifically reference them each time they are used. Next, the following definitions will establish the objects we will study and the concepts to be explored.

Definition 2.3.3. *Let n be an integer.*

(i) *The integer n is **even** if and only if there exists an integer k such that $n = 2k$.*

(ii) *The integer n is **odd** if and only if there exists an integer k such that $n = 2k + 1$.*

For example, the integer 18 is even as $18 = 2 \cdot 9$. The integer 15 is odd as $15 = 2 \cdot 7 + 1$. The integer -21 is also odd as $-21 = 2 \cdot (-11) + 1$. It should be noted that any integer is either even or odd, a fact we will use in subsequent sections and you will prove in Section 2.7. Put another way, if an integer is not even, it is odd, and if it is not odd, it is even.

Definition 2.3.4. *Given two integers n and m, they have the **same parity** if and only if either both n and m are even, or both n and m are odd. If this is not the case, n and m have **opposite parity**.*

As shown above, the integers 15 and -21 have the same parity, while the integers 15 and 18 have opposite parity.

Definition 2.3.5. *Let a and b be integers. Then b **divides** a, denoted $b|a$, if and only if there exists an integer k such that $a = bk$. When this occurs, it is also said that a is a **multiple** of b.*

Note for integers a and b that b does not divide a, denoted $b \nmid a$, if no integer k exists that satisfies $a = bk$. The integer 7 divides 21 or 21 is divisible by 7, denoted by $7|21$, as $21 = 7 \cdot 3$. Along the same lines, $-4|40$ as $40 = (-4) \cdot (-10)$ and $1|5$ as $5 = 1 \cdot 5$. In addition, 4 does not divide 13 or $4 \nmid 13$ as no multiple of 4 results in 13.

Definition 2.3.6. *A positive integer p is **prime** if and only if the only two distinct positive integer divisors of p are 1 and p.*

The integer 2 is the smallest, and only even, prime. All the rest of the primes are odd. There are infinitely many primes, a fact we will prove later on. It should be noted here that if a positive integer is not prime, then it is **composite**. In other words, if n is a composite number, then $n = ab$ for positive integers a and b such that $1 < a, b < n$.

Prime numbers are viewed as the building blocks of the positive integers as indicated by the following result.

Theorem 2.3.7 (Fundamental Theorem of Arithmetic). *Every positive integer greater than 1 is prime or can be written as a product of finitely many primes, with the product being unique up to the order of the primes.*

Now return to the earlier observation that $4 \nmid 13$. We know that the largest multiple of 4 less than or equal to 13 is $12 = 3 \cdot 4$, and that we can add 1 to 12 to get 13. In other words, the integer $13 = 3 \cdot 4 + 1$. The following theorem states the generalized form of this observation.

Theorem 2.3.8 (Division Algorithm). *Given two integers a and b such that $b > 0$, there exists unique integers q and r such that $a = q \cdot b + r$, with $0 \leq r \leq b - 1$.*

The Division Algorithm has a number of handy applications. For example, consider the arbitrary integer a and the positive integer 5. By the Division Algorithm, it follows that $a = 5q + r$ for integers q and r with $0 \leq r \leq 4$. Since there are only five integers from 0 to 4, we know that $a = 5q + 0$, $a = 5q + 1$, $a = 5q + 2$, $a = 5q + 3$, or $a = 5q + 4$. In addition, given integers a and b with $b > 0$, if $a = q \cdot b + 0$ for some integer q, we know that $a = qb$ or that $b|a$. Furthermore, if $a = q \cdot b + r$, where $1 \leq r \leq b - 1$, then we know that b does not divide a or that $b \nmid a$.

Remainders play an important role in mathematics. Consider two integers a and b that have the same remainder when divided by the positive integer n. Then $a = q_1 n + r$ and $b = q_2 n + r$ where q_1, q_2, and r are integers such that $0 \leq r \leq n - 1$. A simple calculation shows that

$$
\begin{aligned}
a - b &= (q_1 n + r) - (q_2 n + r) \\
&= q_1 n - q_2 n + r - r \\
&= (q_1 - q_2) n,
\end{aligned}
$$

where $q_1 - q_2$ is an integer. This means n divides $a - b$, denoted by $n|(a - b)$, motivating the following definition.

Definition 2.3.9. *Let a, b and n be integers such that $n \geq 1$. Then a is **congruent to** b **modulo** n if and only if $n|(a - b)$. This is denoted $a \equiv b \pmod{n}$.*

When two integers a and b are not congruent modulo n, for n a positive integer, the notation $a \not\equiv b \pmod{n}$ is used. Consider the following examples.

- $13 \equiv 3 \pmod 5$: Note $13 - 3 = 10 = 2 \cdot 5$ shows $5|(13 - 3)$.

- $7 \equiv 21 \pmod 2$: Note $7 - 21 = -14 = (-7) \cdot 2$ shows $2|(7 - 21)$.

- $60 \equiv 0 \pmod{10}$: Note $60 - 0 = 60 = 6 \cdot 10$ shows $10|(60 - 0)$.

- $23 \not\equiv 10 \pmod 5$: This follows since $23 - 10 = 13$ and $5 \nmid 13$.

Now for one final observation concerning modulo arithmetic. Using the Division Algorithm with $b = 5$, it follows for any integer a that

$a = 5q+0$, $a = 5q+1$, $a = 5q+2$, $a = 5q+3$, or $a = 5q+4$. This means that $a-0 = 5q$, $a-1 = 5q$, $a-2 = 5q$, $a-3 = 5q$, or $a-4 = 5q$. Given that q is an integer, we have $5|(a-0)$, $5|(a-1)$, $5|(a-2)$, $5|(a-3)$, or $5|(a-4)$. Thus for any integer a, we have that $a \equiv 0 \pmod 5$, $a \equiv 1 \pmod 5$, $a \equiv 2 \pmod 5$, $a \equiv 3 \pmod 5$, or $a \equiv 4 \pmod 5$.

This can be generalized to any integer n with $n \geq 1$. Given an integer a, the Division Algorithm implies $a = qn + r$, for integers q and r such that $0 \leq r \leq n - 1$. This means $a - r = qn$ or that $a \equiv r \pmod n$. This motivates the expression $a \pmod n$, which denotes the remainder when a is divided by n. Consider the following examples.

- $34 \pmod 5 = 4$: This follows since $34 = 6 \cdot 5 + 4$.

- $7 \pmod 2 = 1$: This follows since $7 = 3 \cdot 2 + 1$.

- $30 \pmod{11} = 8$: This follows since $30 = 2 \cdot 11 + 8$.

This sections ends with one final definition. We have made use of it already, but need to formally state the definition now.

Definition 2.3.10. *A number r is* **rational** *if and only if $r = \frac{a}{b}$ for integers a and b such that $b \neq 0$. If r is not rational, then r is* **irrational**.

A rational number $r = \frac{a}{b}$, for integers a and b such that $b \neq 0$, is in **lowest terms** if there is no integer, other than 1 and -1, that divides both a and b.

2.4 DIRECT PROOF

There are numerous results in mathematics of the form $P \rightarrow Q$ or "if P, then Q." The method of **direct proof** to show that a statement in this form is true comes from the truth table for $P \rightarrow Q$.

P	Q	$P \rightarrow Q$
T	T	T
T	F	F
F	T	T
F	F	T

Given that $P \rightarrow Q$ is vacuously true when P is false, there is no need to address this possibility. We must then assume that P is true. The only way for the statement $P \rightarrow Q$ to be true is that Q must then also be true. Thus using the method of direct proof to show that $P \rightarrow Q$

is true involves assuming that P is true, and showing, using a logically sound argument based on the axioms, definitions, and previously proven results, that Q must also be true. Summing up, proving a statement using the method of direct proof will proceed as follows.

Proposition. *If P, then Q.*

> *Proof.* (a) Assume P is true.
> (b) Complete a series of logical deductions.
> (c) Then Q is true. □

Before we begin, there is one final comment on proofs. When writing a proof of a result, you need to let the reader know where the proof begins and ends. All proofs in this text will begin with the word *Proof* in italics (small capitals and bold print are also used), letting the reader know that the proof is beginning. To indicate that a proof is finished, it ends with a symbol right-justified on the same line as the last sentence of the proof. In this text, the symbol □ will be used (the other commonly used symbol is ■).

So let's jump right in. In Section 2.2, we defined a number of mathematical objects and will now prove, using the method of direct proof, a number of results concerning these objects.

Proposition 2.4.1. *If n is an even integer, then $n^2 + 6n - 7$ is odd.*

Proof. Given that n is even, we know by Definition 2.3.3 that $n = 2k$ for some integer k. Then we have

$$n^2 + 6n - 7 = (2k)^2 + 6(2k) - 7$$
$$= 4k^2 + 12k - 8 + 1$$
$$= 2(2k^2 + 6k - 4) + 1,$$

where $2k^2 + 6k - 4$ is an integer. Thus by Definition 2.3.3, it follows that $n^2 + 6n - 7$ is odd. □

As demonstrated in the proof above, the definition of an odd and even integer guided its flow and direction. While Definition 2.3.3 will still be an important aspect in proofs of statements involving parity, from this point on we will stop citing it.

Proposition 2.4.2. *If n is an even integer and m is an odd integer, then $3n + 2m$ is even.*

Proof. Given that n is even and m is odd, there exist integers k and l such that $n = 2k$ and $m = 2l + 1$. A simple calculation shows

$$\begin{aligned} 3n + 2m &= 3(2k) + 2(2l + 1) \\ &= 6k + 4l + 2 \\ &= 2(3k + 2l + 1), \end{aligned}$$

where $3k + 2l + 1$ is an integer. This means that $3n + 2m$ is even. ◻

Now to prove two results concerning division.

Lemma 2.4.3. *Let a, b, and c be integers. If $a|b$ and $b|c$, then $a|c$.*

Proof. Given that $a|b$, by Definition 2.3.5 we know there exists an integer k such that $b = ka$. Similarly, given that $b|c$, there exists an integer l such that $c = lb$. Substituting, we see that

$$c = lb = l(ka) = (lk)a.$$

Since kl is an integer, it follows from Definition 2.3.5 that $a|c$. ◻

In this proof, Definition 2.3.5 guided its construction, but from this point on, while still using this definition, we will stop specifically citing it.

Lemma 2.4.4. *Let a, b, and c be integers such that $a|b$ and $a|c$. Then $a|(bm + cn)$ for any integers m and n.*

Proof. Given that $a|b$ and $a|c$, there exist integers k and l such that $b = ka$ and $c = la$. Substituting, we see that

$$\begin{aligned} bm + cn &= (ka)m + (la)n \\ &= k(am) + l(an) \\ &= k(ma) + l(na) \\ &= (km)a + (ln)a \\ &= (km + ln)a. \end{aligned}$$

Given that $km + la$ is an integer, it follows that $a|(bm + cn)$. ◻

Next we will prove two results concerning modulo arithmetic.

Lemma 2.4.5. *Let a, b, and c be integers, and let n be a positive integer. If $a \equiv b \pmod{n}$, then $ac \equiv bc \pmod{n}$.*

Proof. Given that $a \equiv b \pmod{n}$, by Definition 2.3.9 we know that $n|(a - b)$. This means there exists an integer k such that $a - b = kn$. Multiplying both sides of this equality by c results in $(a - b)c = (kn)c$, which can be rewritten as $ac - bc = k(nc) = k(cn) = (kc)n$. Since kc is an integer, it follows that $n|(ac - bc)$. Thus by Definition 2.3.9, we have that $ac \equiv bc \pmod{n}$. $\qquad\qquad\square$

Again, as demonstrated in the previous proof, the definition of congruence modulo n was central to its construction. While I may sound like a broken record, it can't be stressed enough that proofs are guided by definitions. And as before, from this point on, while still using Definition 2.3.9, we will stop specifically citing it.

Lemma 2.4.6. *Let a and b be integers, and let n be a positive integer. If $a \equiv b \pmod{n}$, then $a^2 \equiv b^2 \pmod{n}$.*

Proof. Given that $a \equiv b \pmod{n}$, we know that $n|(a - b)$. This means there exists an integer k such that $a - b = kn$. Now consider $a^2 - b^2$. By factoring and substituting, we get

$$
\begin{aligned}
a^2 - b^2 &= (a + b)(a - b) \\
&= (a + b)(kn) \\
&= ((a + b)k)n.
\end{aligned}
$$

Given that $(a + b)k$ is an integer, it follows that $n|(a^2 - b^2)$, resulting in $a^2 \equiv b^2 \pmod{n}$. $\qquad\qquad\square$

This section ends with a theorem concerning rational numbers.

Theorem 2.4.7. *The sum, difference, and product of any two rational numbers is rational.*

Before we begin the proof, note that the statement of Theorem 2.4.7 did not introduce any variables. To be able to use the definition of a rational number and create expressions that can be manipulated, this will need to happen in the proof. In the proof below, we will only prove that the sum of two rational numbers is rational as the proofs of the other two statements, being very similar, are exercises.

Proof. Let r and s be rational numbers. By Definition 2.3.10, there exist integers a, b, c, and d, with $b \neq 0$ and $d \neq 0$, such that $r = \frac{a}{b}$ and $s = \frac{c}{d}$. We will show that $r + s$ is also a rational number. To compute $r + s$, a common denominator is needed. Given that $d \neq 0$, we get that $r = \frac{a}{b} = \frac{a}{b} \cdot \frac{d}{d} = \frac{ad}{bd}$. Similarly, since $b \neq 0$, it follows that $s = \frac{c}{d} = \frac{c}{d} \cdot \frac{b}{b} = \frac{bc}{bd}$. A simple calculation shows that

$$r + s = \frac{a}{b} + \frac{c}{d} = \frac{ad}{bd} + \frac{cb}{bd} = \frac{ad + bc}{bd}.$$

First note that both $ab + bc$ and bd are integers. Second, given that $b \neq 0$ and $d \neq 0$, we know that $bd \neq 0$. This means that $r + s$ is rational. □

Exercises 2.4

(1) Prove if n is even and m is odd, then $n + m$ is odd.

(2) Prove if n is odd, then $n^3 + 1$ is even.

(3) Given odd integers n and m, prove for any integer t that $nt + mt$ is even.

(4) Given integer n, prove that if $3n + 2$ is odd, then $5n + 7$ is even.

(5) Let a, b, c, and d be integers. Prove the following statements.

    (a) If $a|b$ and $a|c$, then $a|(b + c)$.   (c) $1|a$ and $a|a$.

    (b) If $a|c$ and $b|d$, then $ab|cd$.   (d) If $a|b$, then $a|(b + a)^2$.

(6) Let a, b, c, d and n be integers such that $n > 0$. Prove the following statements.

    (a) If $a \equiv b \pmod{n}$, then $a + c \equiv b + c \pmod{n}$.

    (b) If $a \equiv b \pmod{n}$, then $a^3 \equiv b^3 \pmod{n}$.

    (c) If $a \equiv b \pmod{n}$ and $c \equiv d \pmod{n}$, then $a + c \equiv b + d \pmod{n}$.

    (d) If $a \equiv b \pmod{n}$ and $c \equiv d \pmod{n}$, then $ac \equiv bd \pmod{n}$.

(7) Prove for $a \in \mathbb{Z}$ that $(a + 1)^3 \equiv a^3 + 1 \pmod{3}$.

(8) Prove for $a \in \mathbb{Z}$ that if $a \equiv 1 \pmod{10}$, then $a^2 \equiv 1 \pmod{10}$.

(9) Let a, b, and n be integers with $n \geq 1$. Prove that if $a \equiv b \pmod{n}$, then a and b have the same remainder when divided by n.

(10) Let a and n be integers such that $n > 0$. Prove that if $a \equiv 1 \pmod{n}$, then $a^2 \equiv 1 \pmod{n}$.

(11) Prove that the difference and product of two rational numbers is rational.

(12) Prove that if r and s are rational, then $\frac{r+s}{2}$ is rational.

(13) Prove that every integer is a rational number.

(14) Prove that if $5r - 2$ is rational, then r is rational.

(15) Prove that if $rs = 1$, where s is a non-zero rational number, then r is rational.

2.5   PROOF BY CONTRAPOSITIVE

Here we will learn a second technique for proving statements of the form $P \to Q$ or "if P, then Q." Using the method of **proof by contrapositive** (also known as **proof by contraposition**) to show that a statement in this form is true comes from the logical equivalence $P \to Q \equiv \sim Q \to \sim P$ (known as the contrapositive) from Theorem 1.3.2. Thus if given the statement $P \to Q$, you can show it is true by showing that $\sim Q \to \sim P$ is true. Summing up, proving a statement using the proof by contrapositive method will proceed as follows.

Proposition. *If P, then Q.*

> *Proof.* (a) Assume $\sim Q$ is true.
> (b) Complete a series of logical deductions.
> (c) Then $\sim P$ is true. □

Proof by contrapositive comes in handy to prove the statement, "if P, then Q" when assuming $\sim Q$ (and showing $\sim P$) is much more straightforward than assuming P (and showing Q). This can occur when it is difficult to get a grasp of P or to meaningfully proceed after assuming P. For example, the statement $a \nmid b$, for integers a and b, may not be easy to work with. But assuming $\sim a \nmid b$ or $a|b$ is readily addressed due to the definition of divides.

Proposition 2.5.1. *Let a be an integer. If $3a - 4$ is an odd integer, then a is odd.*

Proof. We will proceed by using proof by contrapositive. Assume that the integer a is not odd or that a is even. This means that $a = 2k$ for some integer k. Then we have

$$3a - 4 = 3(2k) - 4 = 6k - 4 = 2(3k - 2),$$

where $3k - 2$ is an integer. This implies that $3a - 4$ is even or that $3a - 4$ is not odd. □

It is not required to start a proof of the statement $P \to Q$, where you will use the method of proof by contrapositive, with the statement, "We will proceed by using proof by contrapositive." However, it is also not wrong to do so.

Proposition 2.5.2. *Let a, b, and c be integers. If $a \nmid bc$, then $a \nmid c$.*

Proof. Assume it is not the case that $a \nmid c$ or that $a | c$. Then there exists an integer k such that $c = ka$. Substituting results in

$$bc = b(ka) = (bk)a,$$

where bk is an integer. This implies that $a | bc$ or that it is not the case that $a \nmid bc$. □

Proposition 2.5.3. *Let r be a number. If $\frac{r}{r^2+1}$ is irrational, then r is irrational.*

Proof. Assume that r is not irrational or that r is rational. This means that $r = \frac{k}{l}$ where k and l are integers such that $l \neq 0$. A simple calculation shows

$$\frac{r}{r^2 + 1} = \frac{\frac{k}{l}}{\left(\frac{k}{l}\right)^2 + 1} = \frac{\frac{k}{l}}{\frac{k^2}{l^2} + \frac{l^2}{l^2}} = \frac{\frac{k}{l}}{\frac{k^2+l^2}{l^2}} = \frac{k}{l} \cdot \frac{l^2}{k^2 + l^2} = \frac{kl^2}{l(k^2 + l^2)},$$

where kl^2 and $l(k^2 + l^2)$ are integers. Furthermore, given that $l \neq 0$ and $k^2 \geq 0$, it follows that $k^2 + l^2 \neq 0$. This implies $l(k^2 + l^2) \neq 0$ and that $\frac{r}{r^2+1}$ is rational. As a result, we have that $\frac{r}{r^2+1}$ is not irrational. □

Exercises 2.5

(1) Given integer a, prove that if $5a + 7$ is even, then a is odd.

(2) Given integer a, prove that if $a^2 + 4a - 7$ is odd, then a is even.

(3) Given integer a, prove that if $(a + 1)^3$ is odd, then a is even.

(4) Given integer n, prove that if $5 \nmid n^2$, then $5 \nmid n$.

(5) Given integer a, prove that if $3 \nmid (a + 1)^3$, then $3 \nmid (a^3 + 1)$.

(6) Given integer a, prove that if $4|(a^2 - 1)$, then a is odd.

(7) Given integer a, prove that if $(a + 1)^2 \not\equiv 1 \pmod 3$, then $a \not\equiv 1 \pmod 3$.

(8) Given integer a, prove that if $a^2 \equiv 0 \pmod 5$, then $a \not\equiv 1 \pmod 5$.

(9) Prove that if $r(r + 1)$ is irrational, then r is irrational

(10) Given rational number n, prove that if $n(m + 3)$ is irrational, then m is irrational.

2.6 PROOF BY CASES

There are mathematical results whose conclusion follows from two or more possible hypotheses. For example, consider the following result.

Proposition 2.6.1. *Let n and m be integers. If n and m have the same parity, then $n + m$ is even.*

In Proposition 2.6.1, the hypothesis will be true when both n and m are even and when both n and m are odd. The proof must account for both of these cases. When a situation like this arises, you use the **proof by cases** method. Each case should be made evident to the reader. In this text, each case will begin with "Case (i)" where i refers to the number of the case. Before proceeding, note that different cases may require different proof techniques.

Proof of Proposition 2.6.1. Assume that n and m have the same parity. There are two cases to consider.

Case (1) The integers n and m are both even. In this case, it follows that $n = 2k$ and $m = 2l$ for integers k and l. A quick calculation shows

$$n + m = 2k + 2l = 2(k + l),$$

where $k + l$ is an integer. This means that $n + m$ is even.

Case (2) The integers n and m are both odd. In this case, we know $n = 2k + 1$ and $m = 2l + 1$ for integers k and l. We then have

$$n + m = 2k + 1 + 2l + 1 = 2k + 2l + 2 = 2(k + l + 1),$$

where $k + l + 1$ is an integer. This means that $n + m$ is even.

Each case shows that $n + m$ is always even. $\qquad\square$

Sometimes it takes a little bit of work to determine what the cases are, as demonstrated in the following example.

Proposition 2.6.2. *Let n be an integer. If $3 \nmid n$, then $n^2 \equiv 1 \pmod 3$.*

Proof. Assume that $3 \nmid n$. By the Division Algorithm (Theorem 2.3.8), we know that $n = 3q + r$, for integers q and r such that $0 \le r \le 2$. This implies that $n = 3q + 0, n = 3q + 1$, or that $n = 3q + 2$. If $n = 3q + 0 = 3q$, then $3 | n$. Given that we are assuming that $3 \nmid n$, it must be that $n = 3q + 1$ or $n = 3q + 2$.

Case (1) $n = 3q + 1$. In this case, we have

$$n^2 - 1 = (3q + 1)^2 - 1 = 9q^2 + 6q + 1 - 1 = 3(3q^2 + 2q),$$

where $3q^2 + 2q$ is an integer. This implies that $3 | (n^2 - 1)$ or that $n^2 \equiv 1 \pmod 3$.

Case (2) $n = 3q + 2$. In this case, it follows that

$$
\begin{aligned}
n^2 - 1 &= (3q + 2)^2 - 1 \\
&= 9q^2 + 12q + 4 - 1 \\
&= 9q^2 + 12q + 3 \\
&= 3(3q^2 + 4q + 1),
\end{aligned}
$$

where $3q^2 + 4q + 1$ is an integer. This implies that $3 | (n^2 - 1)$ or that $n^2 \equiv 1 \pmod 3$.

Each case demonstrates that if $3 \nmid n$, then $n^2 \equiv 1 \pmod 3$. ☐

For the following result, we will need De Morgan's Law from Theorem 1.3.2 that states $\sim (P \wedge Q) \equiv \sim P \vee \sim Q$ for statements P and Q.

Proposition 2.6.3. *Let n and m be integers. If $(n + 1)m + 1$ is even, then n is even and m is odd.*

Proof. Assume that it is not the case that n is even and m is odd. This means by De Morgan's Law that n is odd or m is even. There are two cases to examine.

Case (1) n is odd. In this case $n = 2k + 1$ for some integer k. This results in

$$(n + 1)m + 1 = (2k + 1 + 1)m + 1 = (2k + 2)m + 1 = 2(k + 1)m + 1,$$

where $(k + 1)m$ is an integer. This implies that $(n + 1)m + 1$ is odd or that $(n + 1)m + 1$ is not even.

Case (2) m is even. In this case $m = 2l$ for some integer l. This implies

$$(n+1)m + 1 = (n+1)(2l) + 1 = 2(n+1)l + 1,$$

where $(n+1)l$ is an integer. This implies that $(n+1)m + 1$ is odd or that $(n+1)m + 1$ is not even. □

One issue that occasionally comes up when using the proof by cases method is when the proofs of the individual cases are nearly identical. Consider the following example.

Proposition 2.6.4. *Let n and m be integers. If n and m are of opposite parity, then $n + m$ is odd.*

Proof. Assume that n and m are of opposite parity. There are two cases to examine.

Case (1) n is even and m is odd. In this case, $n = 2k$ and $m = 2l + 1$ for integers k and l. A simple calculation shows

$$n + m = 2k + 2l + 1 = 2(k + l) + 1,$$

where $k + l$ is an integer. This means $n + m$ is odd.

Case (2) n is odd and m is even. In this case, $n = 2k + 1$ and $m = 2l$ for integers k and l. This results in

$$n + m = 2k + 1 + 2l = 2k + 2l + 1 = 2(k + l) + 1,$$

where $k + l$ is an integer. This means $n + m$ is odd. □

The calculations used in each case of the proof are essentially the same. The only difference being the role played by n and m. When this occurs, you can roll both cases into one case, letting the reader know you are doing this by saying, "**without loss of generality**." This means that there are two (or more) cases, but each is similar to the one you are about to present so they will be omitted from the proof. The proof of Proposition 2.6.4 can be written as follows.

Proof of Proposition 2.6.4. Assume that n and m are of opposite parity. Without loss of generality, assume that n is even and m is odd. It follows that $n = 2k$ and $m = 2l + 1$ for integers k and l. A quick calculation shows

$$n + m = 2k + 2l + 1 = 2(k + l) + 1,$$

where $k + l$ is an integer. This means $n + m$ is odd. □

We will now move on to proving statements of the form $P \leftrightarrow Q$ for statements P and Q. We know from Theorem 1.3.2 (xii) that the statement $P \leftrightarrow Q$ is logically equivalent to the statement $(P \to Q) \wedge (Q \to P)$. This informs us how to prove statements of the form $P \leftrightarrow Q$. We must prove $P \to Q$ and $Q \to P$. Another way of saying it is that we must prove $P \to Q$ and its converse. In a sense, we have two cases to prove.

When given the statement $P \leftrightarrow Q$, proving $P \to Q$ is referred to as proving the **forward direction**. It is common to label this part of the proof with the symbol $\Rightarrow$. Proving $Q \to P$ is referred to as proving the **reverse direction**. It is common to label this part of the proof with the symbol $\Leftarrow$. Summing up, proving a statement involving a biconditional will proceed as follows.

Proposition. *P if and only if Q.*

Proof. ($\Rightarrow$) Prove $P \to Q$
($\Leftarrow$) Prove $Q \to P$ □

Lemma 2.6.5. *Let n be an integer. Then n is odd if and only if n^2 is odd.*

Proof. To prove n is odd if and only if n^2 is odd, we must show that if n is odd, then n^2 is odd and that if n^2 is odd, then n is odd.

($\Rightarrow$) We must show that if n is odd, then n^2 is odd. Assume that n is odd. Then $n = 2k + 1$ for some integer k. This means

$$n^2 = (2k + 1)^2 = 4k^2 + 4k + 1 = 2(2k^2 + 2k) + 1,$$

where $2k^2 + 2k$ is an integer. This results in n^2 being odd.

($\Leftarrow$) We must show that if n^2 is odd, then n is odd. We proceed using proof by contrapositive and assume that n is not odd or that n is even. Then $n = 2k$ for some integer k. A simple calculation shows

$$n^2 = (2k)^2 = 4k^2 = 2(2k^2),$$

where $2k^2$ is an integer. This yields n^2 is even or that n^2 is not odd. □

There is no need to state at the beginning of the proof, as was done in the previous proof, that to prove $P \leftrightarrow Q$ one must prove $P \to Q$ and $Q \to P$. This will be eliminated from subsequent proofs involving this method. The proof of the following very similar result is an exercise.

Lemma 2.6.6. *Let n be an integer. Then n is even if and only if n^2 is even.*

Proposition 2.6.7. *Let a, b, c, and n be integers such that $n > 0$. Then $a \equiv b \pmod{n}$ if and only of $a + c \equiv b + c \pmod{n}$.*

Proof. ($\Rightarrow$) Assume that $a \equiv b \pmod{n}$. This means that $a - b = qn$ for some integer q. It follows that

$$(a + c) - (b + c) = a + c - b - c$$
$$= a - b$$
$$= qn.$$

Since q is an integer, we have that $a + c \equiv b + c \pmod{n}$.

($\Leftarrow$) Now assume that $a + c \equiv b + c \pmod{n}$. This implies that there exists an integer s such that $(a + c) - (b + c) = sn$. It follows that

$$a - b = a + 0 - b$$
$$= a + c - c - b$$
$$= (a + c) - (b + c)$$
$$= sn.$$

Since s is an integer, we know $a \equiv b \pmod{n}$. □

One final example is given.

Lemma 2.6.8. *Let a be an integer. Then $3 \mid a^2$ if and only if $3 \mid a$.*

Proof. ($\Rightarrow$) To prove that if $3 \mid a^2$, then $3 \mid a$, we will proceed by proof by contrapositive. Assume it is not the case that $3 \mid a$ or that $3 \nmid a$. We know by the Division Algorithm (Theorem 2.3.8) that $a = 3q + r$ for integers q and r such that $0 \leq r \leq 2$. However, if $r = 0$, then $a = 3q + 0 = 3q$ or $3 \mid a$. Since we are assuming that $3 \nmid a$, it must be that $a = 3q + 1$ or $a = 3q + 2$.

Case (1) $a = 3q + 1$. In this case, we get that

$$a^2 = (3q + 1)^2 = 9q^2 + 6q + 1 = 3(3q^2 + 2k) + 1,$$

where $3q^2 + 2q$ is an integer. This means that $3 \nmid a^2$.

Case (2) $a = 3q + 2$. Here, a quick calculation shows that

$$a^2 = (3q + 2)^2$$

$$= 9q^2 + 12q + 4$$
$$= 9q^2 + 12q + 3 + 1$$
$$= 3(3q^2 + 4q + 1) + 1,$$

where $3q^2 + 4q + 1$ is an integer. This also means that $3 \nmid a^2$.

($\Leftarrow$) Assume that $3|a$. Thus there is an integer k such that $a = 3k$. It then follows that

$$a^2 = (3k)^2 = 9k^2 = 3(3k^2),$$

where $3k^2$ is an integer. Thus we have $3|a^2$. □

Exercises 2.6

(1) Given integers a and b, prove that if a and b are of the same parity, then $a^2 + b^2$ is even.

(2) Given integers a and b, prove that if a and b are of opposite parity, then $a^2 + b^2$ is odd.

(3) Given integers $a, b,$ and c, prove that if $a \nmid bc$, then $a \nmid b$ and $a \nmid c$.

(4) Given integers n and m, prove that $4 \nmid (n^2 + m^2 + 1)$.

(5) Given an integer n, prove that $n^2 + n - 2$ is even.

(6) Given an odd integer n, prove that $8|(n-1)(n+1)$.

(7) Given an integer n, prove that $3|(n-1)n(n+1)$.

(8) Prove Lemma 2.6.6: for n an integer, n is even if and only if n^2 is even.

(9) Given an integer a, prove $a^3 + a^2 + a$ is even if and only if a is even.

(10) Given integers n and m, prove nm is odd if and only if n and m are odd.

(11) Given integers n and m, prove nm is even if and only if n is even or m is even.

(12) Given integers a and b, prove $a \equiv b \pmod 5$ if and only if $4a + b \equiv 0 \pmod 5$.

(13) Given an integer a, prove $5|a^2$ if and only if $5|a$.

(14) Prove that 1 and -1 are the only divisors of 1.

2.7 PROOF BY CONTRADICTION

Results in mathematics can take the form of "P" or "$\forall x \in U, P(x)$" for some universe U. The method of **proof by contradiction** starts by assuming the statement $\sim P$ or that there exists an $a \in U$ such that $\sim P(a)$. Based on this assumption, you make a series of logical deductions that result in a contradiction C, a statement that is always false. This results in $\sim P \to C$ or $\sim P(a) \to C$ being a true statement. But given that C is false, the only way for $\sim P \to C$ or $\sim P(a) \to C$ to be true is if $\sim P$ or $\sim P(a)$ is false. In the first case, since $\sim P$ is false, it follows that P is true. In the second, since $\sim P(a)$ is false, where a could have been any value in U, then for all x in U, the statement $P(x)$ must be true. Summing up, proving the statement of the form of "P" or "$\forall x \in S, P(x)$" using the method of proof by contradiction will proceed as follows.

Proposition 2.7.1. P

> *Proof.* (a) Assume $\sim P$
> (b) Complete a series of logical deductions.
> (c) A contradiction C is obtained
> (d) Conclude P is true □

We begin by proving the following well-known result.

Theorem 2.7.2. *The number $\sqrt{2}$ is irrational.*

Proof. Suppose that $\sqrt{2}$ is rational. This means that $\sqrt{2} = \frac{a}{b}$ for integers a and b, with $b \neq 0$ and $\frac{a}{b}$ in lowest terms. This implies that

$$\left(\frac{a}{b}\right)^2 = \left(\sqrt{2}\right)^2$$

or that

$$\frac{a^2}{b^2} = 2.$$

Multiplying both sides by b^2 results in $a^2 = 2b^2$. Given that b^2 is an integer, this implies that a^2 is even. By Lemma 2.6.6, we know that if a^2 is even, then a is even. Therefore, $a = 2k$ for some integer k.

Proceed by substituting $a = 2k$ into the equation $a^2 = 2b^2$, which results in

$$(2k)^2 = 2b^2$$

or

$$4k^2 = 2b^2.$$

This means that $2(2k^2) = 2b^2$ or that $2k^2 = b^2$. Given that k^2 is an integer, the fact that $b^2 = 2k^2$ implies that b^2 is even. Applying Lemma 2.6.6 again results in the fact that b is even. Therefore, $b = 2l$ for some integer l. Therefore, $2|a$ and $2|b$, which is a contradiction as $\frac{a}{b}$ is in lowest terms. This contradiction implies that $\sqrt{2}$ is irrational. □

Proposition 2.7.3. *For a and b even integers, $a^2 - b^2 \neq 2$.*

Proof. Suppose there exist even integers a and b such that $a^2 - b^2 = 2$. Given that a and b are even, there exist integers k and l such that $a = 2k$ and $b = 2l$. Given that $a^2 - b^2 = 2$ we have that

$$(2k)^2 - (2l)^2 = 2$$

or that

$$4k^2 - 4l^2 = 2.$$

This means $2(2k^2 - 2l^2) = 2 \cdot 1$ or that $2k^2 - 2l^2 = 1$. We know that 1 is odd ($1 = 2 \cdot 0 + 1$). However, $2k^2 - 2l^2 = 2(k^2 - l^2)$, where $k^2 - l^2$ is an integer, implying that $2k^2 - 2l^2$ or 1 is even. This is a contradiction. This means that for all even integers a and b, $a^2 - b^2 \neq 2$. □

Given statements P and Q, you can also prove statements of the form $P \to Q$ or "if P, then Q" using proof by contradiction. The method is based on the logical equivalence $\sim (P \to Q) \equiv P \land \sim Q$ from Theorem 1.3.2 (xiii). It starts by assuming $\sim (P \to Q)$, which is equivalent to assuming $P \land \sim Q$. You then construct an argument that results in a contradiction C, a statement that is always false. This results in $(P \land \sim Q) \to C$ being a true statement. But given that C is always false, the only way for $(P \land \sim Q) \to C$ to be true is if $P \land \sim Q$ is false. This means that $\sim (P \to Q)$ must be false or that $P \to Q$ is true. This technique is very similar to proof by contrapositive.

Proposition 2.7.4. *Let n be an integer. If $n^2 + 4n + 5$ is even, then n is odd.*

Proof. Assume that $n^2 + 4n + 5$ is even and that n is not odd. Thus n is even and there exists an integer k such that $n = 2k$. This results in

$$n^2 + 4n + 5 = (2k)^2 + 4(2k) + 5$$

$$= 4k^2 + 8k + 4 + 1$$
$$= 2(2k^2 + 4k + 2) + 1,$$

where $2k^2 + 4k + 2$ is an integer. This means that $n^2 + 4n + 5$ is odd, which is a contradiction to the assumption that $n^2 + 4n + 5$ is even. This implies that if $n^2 + 4n + 5$ is even, then n is odd. □

Exercises 2.7

(1) Prove that $\sqrt{3}$ is irrational.

(2) Prove that an integer can't be both even and odd.

(3) Prove for all odd integers a and b that $4 \nmid (a^2 + b^2)$.

(4) Prove for all odd integers a and even integers b that $a^2 - b^2 \neq 0$.

(5) Prove for all integers a and b that $8a + 4b \neq 7$.

(6) Prove for all integers a that $5 \nmid a$ or $5 \nmid a + 1$.

(7) Prove for all integers n that if $n^2 + 6n + 1$ is odd, then n is even.

(8) Given an integer a such that $a \geq 2$, prove that if $a|b$, then $a \nmid (b+1)$.

(9) Given integers a and n such that $n > 0$, prove that if $a \not\equiv 1 \pmod{n}$, then $a^2 \not\equiv 1 \pmod{n}$.

(10) Given integers a and b, prove that if $a^3 \not\equiv b^3 \pmod{3}$, then $(a - b)^3 \not\equiv 0 \pmod{3}$.

CHAPTER 3

Sets

3.1 THE CONCEPT OF A SET

We have alluded to the concept of a set in Chapter 1 and Chapter 2 where specific collections of numbers, such as the integers $\mathbb{Z}$, were used. Our study of set theory will take the naive approach where the term set will not formally be defined.

We will understand a **set** to be a collection of distinct objects, called **elements** of the set, where the objects contained in the set are evident. Capital letters $A, B, C, \ldots$ are used to denote sets with lower case letters $a, b, c, \ldots$ used to denote elements of sets. If an element a is an object in or member of the set A, this is denoted by $a \in A$. If an object b is not in the set A, this is denoted by $b \notin A$. The symbol $\{$ is used to denote where the presentation of the elements in the set begins and the symbol $\}$ is used to denote when this presentation ends.

There are a number of ways to present a set. For example, consider the following set

$$A = \{1, 2, 3, 4, \alpha, \beta, \gamma, \delta\}.$$

Here the elements of the set are written out in a list. It is clear that $1 \in A$ and that $6 \notin A$. It should be noted that the order in which the elements are listed does not matter. However, writing the elements of a set in a universally recognized order can help to more quickly determine the elements in the set. If you are unfamiliar with them, the symbols α (lower-case alpha), β (lower-case beta), γ (lower-case gamma), and δ (lower-case delta) are the first four letters, respectively, of the Greek alphabet.

When explicitly listing all of the elements of a set becomes too unwieldy, the ellipsis symbol $\ldots$ is used. This symbol indicates that the pattern that has been established should continue until an end is reached or should continue forever. For example, the set

$$B = \{1, 2, 3, \ldots, 50\}$$

is the collection of integers from 1 to 50. We know that $36 \in B$, but that $51 \notin B$. We saw the use of the ellipsis symbol earlier when we introduced the natural numbers $\mathbb{N}$ where

$$\mathbb{N} = \{1, 2, 3, 4, \ldots\}.$$

Here the pattern goes on forever. Another example is the set

$$C = \left\{1, \frac{1}{2}, \frac{1}{2^2}, \frac{1}{2^3}, \ldots\right\}.$$

The ellipsis symbol can be used to indicate a continuing pattern in two different directions. We saw this earlier when we introduced the integers $\mathbb{Z}$ as

$$\mathbb{Z} = \{\ldots, -3, -2, -1, 0, 1, 2, 3, \ldots\}.$$

While we will predominately deal with sets of numbers, a set can be any collection of objects. For example, the set

$$D = \{\text{red, orange, yellow, green blue, indigo, violet}\}$$

is the set of colors of a rainbow. For another example, consider the set

$$E = \{2, \{3\}, \{2, \{3\}\}\}.$$

The number 2 is an element of this set ($2 \in E$), but 3 is not in the set ($3 \notin E$). However, the sets $\{3\}$ and $\{2, \{3\}\}$ are elements of E.

A set is called a **finite set** if it contains a finite number of elements. If a set is not finite, then it is an **infinite set**. The sets $A, B,$ and D listed above are finite sets, while the sets $C, \mathbb{N},$ and $\mathbb{Z}$ are infinite sets. A much deeper investigation of these concepts will be conducted in Chapter 7.

A very important and fundamental set is the **empty set**, denoted by $\emptyset$ or $\{\}$, which is a finite set containing no or zero elements. Be careful, the set $\{\emptyset\}$ is not the empty set. It is a set with one element in it, namely the empty set ($\emptyset \in \{\emptyset\}$).

In addition to listing the elements of a set, either completely or using the ellipsis symbol, **set-builder notation** can be used to determine a set. There are two ways in which this notation is used to describe a set S. The first comes in the form

$$S = \{f(x) : x \in U\},$$

where $f(x)$ is some expression involving x, and U is the collection of objects from which x can be chosen. For example, consider the set

$$F = \left\{\frac{1}{n} : n \in \mathbb{N}\right\}.$$

Here, set-builder notation has been used to determine the set $F = \{1, \frac{1}{2}, \frac{1}{3}, \frac{1}{4}, \ldots\}$. The classic example of a set defined this way, as we have seen earlier, is the rational numbers $\mathbb{Q}$ where

$$\mathbb{Q} = \left\{\frac{a}{b} : a, b \in \mathbb{Z} \text{ with } b \neq 0\right\}.$$

The other method is of the form

$$S = \{x \in U : P(x)\},$$

where U is a collection of objects and $P(x)$ is a predicate involving x. An object $a \in U$ is in set S if $P(a)$ is true and it is not an element of set S if $P(a)$ is false. For example, the set

$$G = \{n \in \mathbb{Z} : |n| \geq 5\}$$

is the collection in integers $\{\ldots, -7, -6, -5, 5, 6, 7, \ldots\}$. For another example, the set

$$H = \{x \in \mathbb{R} : x^3 - 3x^2 = 0\}$$

is the collection of real numbers $\{-\sqrt{3}, 0, \sqrt{3}\}$.

Now for a definition.

Definition 3.1.1. *The **ordered pair** of two objects x and y is the ordered list (x, y).*

Given two distinct objects x and y, it follows that $\{x, y\} = \{y, x\}$ but that $(x, y) \neq (y, x)$ as the pairs are ordered. For the ordered pairs (x_1, x_2) and (y_1, y_2), we have $(x_1, x_2) = (y_1, y_2)$ if and only if $x_1 = y_1$ and $x_2 = y_2$. The concept of an ordered pair leads to the next definition.

Definition 3.1.2. *Given two sets A and B, the **Cartesian product** of A and B, denoted by $A \times B$, is the set $\{(a, b) : a \in A, b \in B\}$.*

For example, given $A = \{1, 2, 3\}$ and $B = \{\alpha, \beta\}$, then

$A \times B = \{(1, \alpha), (2, \alpha), (3, \alpha), (1, \beta), (2, \beta), (3, \beta)\}$,
$B \times A = \{(\alpha, 1), (\alpha, 2), (\alpha, 3), (\beta, 1), (\beta, 2), (\beta, 3)\}$, and
$A \times A = \{(1, 1), (1, 2), (1, 3), (2, 1), (2, 2), (3, 2), (3, 1), (3, 2), (3, 3)\}$.

As highlighted by plotting points in three-dimensional space, where you plotted ordered triples (x, y, z), there is interest in ordered lists of more than two elements.

Definition 3.1.3. *The **ordered n-tuple** of the objects $x_1, x_2, \ldots, x_n$, where n is an integer such that $n \geq 2$, is the ordered list $(x_1, x_2, \ldots, x_n)$.*

Ordered 2-tuples are more simply referred to as ordered pairs, and ordered 3-tuples are more commonly referred to as ordered triples. Given two ordered n-tuples $(x_1, x_2, \ldots, x_n)$ and $(y_1, y_2, \ldots, y_n)$, we have $(x_1, x_2, \ldots, x_n) = (y_1, y_2, \ldots, y_n)$ if and only if $x_1 = y_1, x_2 = y_2, \ldots, x_n = y_n$ ($x_i = y_i$ for $1 \leq i \leq n$).

Definition 3.1.4. *Given sets $A_1, A_2, \ldots, A_n$, for n an integer with $n \geq 2$, the **Cartesian product of the sets** $A_1, A_2, \ldots, A_n$, denoted by $A_1 \times A_2 \times \cdots \times A_n$, is the set of ordered n-tuples $\{(a_1, a_2, \ldots, a_n) : a_1 \in A_1, a_2 \in A_2, \ldots, a_n \in A_n\}$.*

For example, $(100, 3, \sqrt{2}, -20)$ would be an element of the Cartesian product $\mathbb{Q} \times \{1, 2, 3\} \times \mathbb{R} \times \mathbb{Z}$. If n is a positive integer with $n \geq 2$, then the Cartesian product

$$A^n = \underbrace{A \times A \times \cdots \times A}_{n\text{-times}} = \{(a_1, a_2, \ldots, a_n) : a_i \in A \text{ for } 1 \leq i \leq n\}.$$

Exercises 3.1

(1) Determine if the following statements are true or false. Be sure to support your answer.

(a) $2 \in \{0, 2, 4, 6, 8\}$

(b) $4 \notin \{0, 2, 4, 6, 8\}$

(c) $2 \in \{0, 1, \{2\}\}$

(d) $\{2\} \in \{0, 1, \{2\}\}$

(e) $\{3\} \in \{\{3\}\}$

(f) $\emptyset \in \{-1, 0, 1\}$

(g) $\emptyset \in \{-1, \emptyset, 1\}$

(h) $\emptyset \in \{-1, \{\emptyset\}, 1\}$

(2) List the elements in each of the following sets.

(a) $\{n \in \mathbb{Z} : \frac{6}{n} \in \mathbb{Z}\}$

(b) $\{x \in \mathbb{R} : x^2 + x - 6 = 0\}$

(c) $\{r \in \mathbb{Q} : \frac{1}{r} = r\}$

(d) $\{n \in \mathbb{N} : n \text{ is prime}\}$

(e) $\{x \in \mathbb{R} : \frac{1}{x^2} = 0\}$

(f) $\{r \in \mathbb{Q} : r = 2^n \text{ for } n \in \mathbb{Z}\}$

(g) $\{n \in \mathbb{Z} : n \equiv 1 \pmod 4\}$

(h) $\{(a, b) \in \mathbb{Z}^2 : a^2 + b^2 \leq 2\}$

(i) $\{(a, b, c) \in \mathbb{N}^3 : a+b+c \leq 4\}$

(j) $\{n \in \mathbb{N} : 5|n\}$

(k) $\{n \in \mathbb{N} : \frac{1}{n} > 1\}$

(l) $\{3n + 1 : n \in \mathbb{Z}\}$

(m) $\{n^2 : n \in \mathbb{N}\}$

(n) $\{2^n : n \in \mathbb{Z}\}$

(3) Write each of the following sets using set-builder notation.

(a) $\{1, 2, 3, 4, 5\}$

(b) $\{\ldots, -5, -3, -1, 1, 3, 5, \ldots\}$

(c) $\{\ldots, -24, -18, -12, -6, 0, 6, 12, 18, 24, \ldots\}$

(d) $\{\ldots, \frac{1}{27}, \frac{1}{9}, \frac{1}{3}, 1, 3, 9, 27, \ldots\}$

(e) $\{2, 3, 5, 7, 11, 13, 17, 19, 23, \ldots\}$

(f) $\{1, 2, 5, 10, 17 \ldots\}$

(g) $\{\ldots, -18, -13, -8, -3, 2, 7, 12, 17, 22, \ldots\}$

(h) $\{\ldots, -\sqrt{5}, -2, -\sqrt{3}, -\sqrt{2}, -1, 1, \sqrt{2}, \sqrt{3}, 2, \sqrt{5}, \ldots\}$

(i) $\{\ldots, (-2, -2), (-1, -1), (0, 0), (1, 1), (2, 2), \ldots\}$

(j) $\{\ldots, (-3, 9), (-2, 4), (-1, 1), (0, 0), (1, 1), (2, 4), (3, 9), \ldots\}$

(k) $\{(3, 4, 5), (5, 12, 13), (6, 8, 10), (7, 24, 25), \ldots\}$

(4) For sets $A = \{0, 1\}, B = \{2, 4, 6, 8\}$, and $C = \{i, -1\}$, list the elements of the following sets.

(a) $A \times B$

(b) $B \times A$

(c) $\{n \in \mathbb{Z} : |n| < 3\} \times C$

(d) $A \times B \times C$

(e) $A \times (C \times A)$

(f) $A \times \emptyset$

(g) $C \times C$

(h) A^3

(5) Determine the follow sets.

(a) $\{n \in \mathbb{N} : 3n < 12\} \times \{\frac{1}{2^m} : m \in \{-1, 0, 1\}\}$

(b) $\{x \in \mathbb{R} : x^2 = 5\} \times \{r \in \mathbb{Q} : r^2 = \frac{1}{4}\} \times \{n \in \mathbb{Z} : n^2 = 4\}$

(c) $\{\emptyset\} \times \{\emptyset, \{\emptyset\}\} \times \{\emptyset, \{\emptyset\}, \{\emptyset, \{\emptyset\}\}\}$

3.2 SUBSETS AND SET EQUALITY

Definition 3.2.1. *Given sets A and B, the set A is a **subset** of set B, denoted by $A \subseteq B$, if and only if every element of A is also an element of B.*

If a set A is not a subset of a set B, this is denoted by $A \nsubseteq B$. From the definition it follows that

$$\{1,2,3\} \subsetneq \{0,1,2,3,4,5\}$$

and

$$\{1,2,3\} \subseteq \{1,2,3\}.$$

These two examples highlight two important ideas. In the first example, there were elements in the set $\{0,1,2,3,4,5\}$ that are not in the set $\{1,2,3\}$. In general, when $A \subseteq B$ for sets A and B with there being elements in B that are not in A, we say that A is a **proper subset** of B and denote this by $A \subset B$. In our example, we have $\{1,2,3\} \subset \{0,1,2,3,4,5\}$. Using this language, the set $\{1,2,3\}$ is not a proper subset of the set $\{1,2,3\}$. In fact, they are the same set. The second example above motivates the fact that $A \subseteq A$ for any set A.

Given the set of real numbers $\mathbb{R}$, there are eight common proper subsets of $\mathbb{R}$ or intervals that will appear regularly in this book. For $a, b \in \mathbb{R}$, the intervals are defined as follows:

- $(a,b) = \{x \in \mathbb{R} : a < x < b\}$
- $[a,b) = \{x \in \mathbb{R} : a \leq x < b\}$
- $(a,b] = \{x \in \mathbb{R} : a < x \leq b\}$
- $[a,b] = \{x \in \mathbb{R} : a \leq x \leq b\}$
- $(a,\infty) = \{x \in \mathbb{R} : a < x\}$
- $[a,\infty) = \{x \in \mathbb{R} : a \leq x\}$
- $(-\infty,b) = \{x \in \mathbb{R} : x < b\}$
- $(-\infty,b] = \{x \in \mathbb{R} : x \leq b\}$

In addition, when given the set of integers $\mathbb{Z}$, the set of rational numbers $\mathbb{Q}$, or the set of real numbers $\mathbb{R}$, the superscript $^+$ denotes the positive numbers in that set, the superscript $^-$ denotes the negative numbers in that set, and the superscript * denotes the non-zero numbers in that set. For example, $\mathbb{Z}^-$ denotes the negative integers, $\mathbb{Q}^+$ denotes the positive rational numbers, and $\mathbb{R}^*$ denotes the non-zero real numbers.

A quick comment on notation before we continue. Don't confuse or misuse the symbols $\in$ and $\subseteq$. For example, consider the set $S = \{1, \{2\}, \{1,2\}\}$. The object 1 is an element of S ($1 \in S$), but is not a set itself, so it can't be a subset of S ($1 \not\subseteq S$). The object $\{2\}$ is a set and also is an element of S ($\{2\} \in S$), whereas 2 is neither a set nor an element of S. But $\{\{2\}\}$ is a set with the element $\{2\}$ in it, which is an element of S, so it is a subset of S ($\{\{2\}\} \subseteq S$). The same line of reasoning shows $\{1,2\} \in S$, $\{1,2\} \not\subseteq S$, and that $\{\{1,2\}\} \subseteq S$.

Lemma 3.2.2. *Let A, B, and C be sets. Then*

(i) $A \subseteq A$,

(ii) $\emptyset \subseteq A$, and

(iii) *if $A \subseteq B$ and $B \subseteq C$, then $A \subseteq C$.*

Only the first two statements from Lemma 3.2.2 will be proven. The last is an exercise. The definition of subsets motivates how one proves that $X \subseteq Y$ for sets X and Y. Start by letting $x \in X$. If you can show that $x \in Y$, then every element of X is also an element of Y and $X \subseteq Y$.

Proof of Lemma 3.2.2 (i) and (ii). To prove (i), let $a \in A$. Given that this means $a \in A$, we have that $A \subseteq A$.

For (ii), start by letting $x \in \emptyset$, or stated another way, by letting x be an element of the empty set. This statement is false as the empty set is the set with no elements. Thus regardless of whether $x \in A$ is true or false, the implication if $x \in \emptyset$, then $x \in A$ is vacuously true. This means $\emptyset \subseteq A$. □

So what does it mean for two sets A and B to be equal? Given our agreed upon notion of a set, it makes sense that they are equal if they contain the same elements. Given Lemma 3.2.2 (i), it then follows that if A and B are the same set, then $A \subseteq B$ and $B \subseteq A$. This motivates the following definition.

Definition 3.2.3. *Let A and B be sets. Then the two sets are **equal**, denoted $A = B$, if and only if $A \subseteq B$ and $B \subseteq A$.*

Definition 3.2.3 provides us a way to show that two sets are equal. Equality, with regards to sets, has the identity, commutative, and transitive properties you would expect, as shown in the next lemma.

Lemma 3.2.4. *Let A, B, and C bet sets. Then*

(i) $A = A$,

(ii) *if $A = B$, then $B = A$, and*

(iii) *if $A = B$ and $B = C$, then $A = C$.*

We will only prove (iii) from Lemma 3.2.4 with the proofs of the other two properties being exercises.

Proof of Lemma 3.2.4 (iii). To show $A \subseteq C$, let $x \in A$. Given that $A = B$, we know that $A \subseteq B$. This implies $x \in B$. Since $B = C$, we are also given that $B \subseteq C$. Thus $x \in B$ implies $x \in C$. We now have $A \subseteq C$.

To show that $C \subseteq A$, let $y \in C$. Since we are given that $B = C$, we know that $C \subseteq B$. Thus $y \in C$ implies that $y \in B$. Similarly, since we are given that $A = B$, we know $B \subseteq A$. Since $y \in B$, we have that $y \in A$. As a result, $C \subseteq A$. □

Now onto the final concept to be introduced in this section.

Definition 3.2.5. *Given a set A, the **power set** of A, denoted by $\mathcal{P}(A)$, is the set of all subsets of A, that is*

$$\mathcal{P}(A) = \{X : X \subseteq A\}.$$

The power set of the sets $\emptyset$, $\{a\}$, $\{a, b\}$, and $\{a, b, c\}$ are formed below.

Set	Subsets	No. Subsets
$\emptyset$	$\emptyset$	1
$\{a\}$	$\emptyset, \{a\}$	2
$\{a, b\}$	$\emptyset, \{a\}, \{b\}, \{a, b\}$	4
$\{a, b, c\}$	$\emptyset, \{a\}, \{b\}, \{c\}, \{a, b\}, \{a, c\}, \{b, c\}, \{a, b, c\}$	8

This results in $\mathcal{P}(\emptyset) = \{\emptyset\}$, $\mathcal{P}(\{a\}) = \{\emptyset, \{a\}\}$, $\mathcal{P}(\{a, b\}) = \{\emptyset, \{a\}, \{b\}, \{a, b\}\}$, and $\mathcal{P}(\{a, b, c\}) = \{\emptyset, \{a\}, \{b\}, \{c\}, \{a, b\}, \{a, c\}, \{b, c\}, \{a, b, c\}\}$. You may have also noticed a relation between the number of elements in a set and the number of elements in its power set. We will talk about this more in Chapter 7.

Example 3.2.6.

(i) $\mathcal{P}(\{1, \{3\}\}) = \{\emptyset, \{1\}, \{\{3\}\}, \{1, \{3\}\}\}$

(ii) $\mathcal{P}(\{1\} \times \{2, 3\}) = \mathcal{P}(\{(1, 2), (1, 3)\})$
$$= \{\emptyset, \{(1, 2)\}, \{(1, 3)\}, \{(1, 2), (1, 3)\}\}$$

(iii) $\mathcal{P}(\mathcal{P}(\{1\})) = \mathcal{P}(\{\emptyset, \{1\}\}) = \{\emptyset, \{\emptyset\}, \{\{1\}\}, \{\emptyset, \{1\}\}\}$

You can create the power set of infinite sets as well, such as $\mathcal{P}(\mathbb{N})$, the set of all subset of $\mathbb{N}$, and $\mathcal{P}(\mathbb{R})$, the set of all subsets of $\mathbb{R}$. For example, $\{1, 2, 4, 8, 16, \ldots\} \in \mathcal{P}(\mathbb{N})$ and $\{-1, 0, e\} \in \mathcal{P}(\mathbb{R})$.

Exercises 3.2

(1) Determine if the following statements are true or false. Be sure to support your answer.

(a) $3 \in \{2, \{3\}\}$ (i) $\emptyset \subseteq \mathbb{R}$

(b) $3 \subseteq \{2, \{3\}\}$ (j) $\mathbb{R} \subseteq \{\mathbb{R}\}$

(c) $2 \in \{2, \{3\}\}$ (k) $\mathbb{R} \in \{\mathbb{R}\}$

(d) $\{3\} \subseteq \{1, \{3\}\}$ (l) $1 \in \mathcal{P}(\{1, 2, 3\})$

(e) $\{2, 3\} \subseteq \{1, 2, \{2, 3\}\}$ (m) $\{1\} \in \mathcal{P}(\{1, 2, 3\})$

(f) $\{2, 3\} \in \{1, 2, \{2, 3\}\}$ (n) $\emptyset \in \mathcal{P}(\{1, 2, 3\})$

(g) $\emptyset \in \{1, 2, 3\}$ (o) $\{\emptyset\} \in \mathcal{P}(\{1, 2, 3\})$

(h) $\emptyset \subseteq \{1, 2, 3\}$ (p) $\{\emptyset\} \in \mathcal{P}(\{\emptyset, 1\}\})$

(2) Determine the indicated sets.

(a) $\mathcal{P}(\{a, b, c, d\})$ (c) $\mathcal{P}(\{a\}) \times \{b, c\}$

(b) $\mathcal{P}(\mathcal{P}(\emptyset))$ (d) $\mathcal{P}(\{a\}) \times \mathcal{P}(\{b, c\})$

(3) Given an example of a set A and an element a such that $a \in A$ and $a \subseteq A$.

(4) Given a set A, prove that if $\{a, b\} \in \mathcal{P}(A)$, then $a \in A$ and $b \in A$.

(5) Let A and B be sets. Prove that $A = \emptyset$ or $B = \emptyset$ if and only if $A \times B = \emptyset$.

(6) Prove statement Lemma 3.2.2 (iii): for sets A, B and C, if $A \subseteq B$ and $B \subseteq C$, then $A \subseteq C$.

(7) Prove statement Lemma 3.2.4 (i): for A a set, $A = A$.

(8) Prove statement Lemma 3.2.4 (ii): for sets A and B, if $A = B$, then $B = A$.

(9) Given sets A, B, C, and D, prove that if $A \subseteq B$ and $C \subseteq D$, then $A \times C \subseteq B \times D$.

(10) Given sets A, B, and C, with $C \neq \emptyset$, prove that if $A \times C = B \times C$, then $A = B$.

(11) Given sets A and B such that $A \subseteq B$, prove that $\mathcal{P}(A) \subseteq \mathcal{P}(B)$.

(12) Given sets A, B, and C, prove that if $A \subseteq B, B \subseteq C$, and $C \subseteq A$, then $A = C$.

3.3 OPERATIONS ON SETS

It is now time to define three fundamental set operations.

Definition 3.3.1. *Let A and B be sets.*

(i) *The **union** of A and B is the set $A \cup B = \{x : x \in A \text{ or } x \in B\}$.*

(ii) *The **intersection** of A and B is the set $A \cap B = \{x : x \in A \text{ and } x \in B\}$.*

(iii) *The **difference** of A and B is the set $A - B = \{x : x \in A \text{ and } x \notin B\}$.*

These, and other, set operations can be visualized with a **Venn diagram**. In a Venn diagram, each set is represented as a circle and the shaded region in the diagram corresponds to the elements satisfying a specific operation. The Venn diagrams for $A \cup B$, $A \cap B$, and $A - B$, for sets A and B are given in Figure 3.1.

Example 3.3.2. *For the sets $A = \{1, 2, 3, 4, 5\}, B = \{0, 1, 3, 7\}$, and $C = \{8, 9, 10\}$, the following sets are constructed. Parentheses are used to clear up any ambiguities.*

- $A \cup B = \{0, 1, 2, 3, 4, 5, 7\}$

- $A \cap B = \{1, 3\}$

- $A \cap C = \emptyset$

- $A - B = \{2, 4, 5\}$

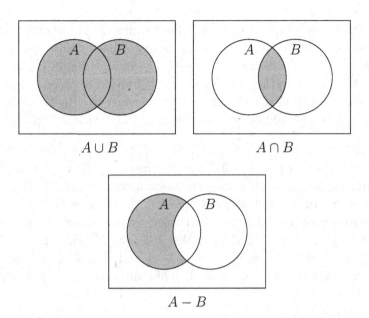

Figure 3.1 Venn diagrams for the union, intersection, and difference of the sets A and B.

- $B - A = \{0, 7\}$
- $B - C = \{0, 1, 3, 7\} = B$
- $(A - B) \cup (B - A) = \{2, 4, 5\} \cup \{0, 7\} = \{0, 2, 4, 5, 7\}$
- $(A \cap B) \times C = \{1, 3\} \times \{8, 9, 10\}$
  $$= \{(1, 8), (1, 9), (1, 10), (3, 8), (3, 9), (3, 10)\}$$
- $(A \times C) \cap (B \times C) = \{(1, 8), (3, 8), (1, 9), (3, 9), (1, 10), (3, 10)\}$

As with most operations in mathematics, set operations have a number of properties. We will start with the most basic.

Lemma 3.3.3. *Let A be a set.*

(i)	$A \cap \emptyset = \emptyset$	*Identity Law*
(ii)	$A \cup \emptyset = A$	*Identity Law*
(iii)	$A \cap A = A$	*Idempotent Law*
(iv)	$A \cup A = A$	*Idempotent Law*

We will only prove (i) and (iv) from Lemma 3.3.3; the other two are exercises. But before proceeding, we need to discuss an application of Theorem 1.3.1 (ii), which states $P \rightarrow P \lor Q$ for statements P and Q. Let X and Y be sets, and let $x \in X$. By the definition of union, an element is in $X \cup Y$ when that element is in X or when it is in Y. Given that $x \in X$, it directly follows from Theorem 1.3.1 (ii) that $x \in X \cup Y$.

Proof of Lemma 3.3.3 (i) and (iv). For (i), to show $A \cap \emptyset \subseteq \emptyset$, let $x \in A \cap \emptyset$. Then $x \in A$ and $x \in \emptyset$. This statement is false as the empty set contains no elements. Therefore, the statement if $x \in A \cap \emptyset$, then $x \in \emptyset$ is vacuously true. It follows that $A \cap \emptyset \subseteq \emptyset$. The fact that $\emptyset \subseteq A \cap \emptyset$ follows from Lemma 3.2.2 (ii) (the empty set is a subset of any set).

For (iv), to show $A \cup A \subseteq A$, let $x \in A \cup A$. This means $x \in A$ or $x \in A$. In either case, we have $x \in A$ and $A \cup A \subseteq A$ follows. To show that $A \subseteq A \cup A$, let $x \in A$. This implies $x \in A \cup A$, yielding $A \subseteq A \cup A$. $\qquad\square$

There are a number of properties that set operations satisfy.

Theorem 3.3.4. *Let $A, B,$ and C be sets.*

(i)	$A \cup B = B \cup A$	*Commutative Law*
(ii)	$A \cap B = B \cap A$	*Commutative Law*
(iii)	$A \cup (B \cup C) = (A \cup B) \cup C$	*Associative Law*
(iv)	$A \cap (B \cap C) = (A \cap B) \cap C$	*Associative Law*
(v)	$A \cup (B \cap C) = (A \cup B) \cap (A \cup C)$	*Distributive Law*
(vi)	$A \cap (B \cup C) = (A \cap B) \cup (A \cap C)$	*Distributive Law*
(vii)	$A - (B \cup C) = (A - B) \cap (A - C)$	
(viii)	$A - (B \cap C) = (A - B) \cup (A - C)$	
(ix)	$A \times (B \cup C) = (A \times B) \cup (A \times C)$	
(x)	$A \times (B \cap C) = (A \times B) \cap (A \times C)$	

Before proving (vi), (vii), and (ix) from Theorem 3.3.4, two observations need to be made. The first is that Venn diagrams can be used to provide evidence, not prove, that an identity concerning sets is true. We will use Venn diagrams to support the fact that the Distributive Law

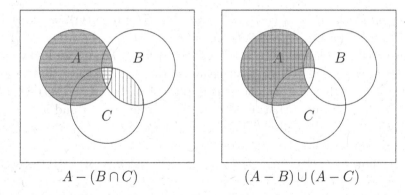

$$A - (B \cap C) \qquad (A - B) \cup (A - C)$$

Figure 3.2 Venn diagrams for sets $A - (B \cap C)$ and $(A - B) \cup (A - C)$.

$A - (B \cap C) = (A - B) \cup (A - C)$ is true. In the Venn diagram on the left in Figure 3.2, the region denoting set A is shaded with horizontal stripes and the region denoting $B \cap C$ is shaded with vertical stripes. Removing the vertical striped region from the horizontally striped region results in the gray shaded region representing $A - (B \cap C)$.

In the Venn diagram on the right in Figure 3.2, the region denoting set $A - B$ is shaded with horizontal stripes and the region denoting $A - C$ is shaded with vertical stripes. Combining the two striped regions results in the gray shaded region representing $(A - B) \cup (A - C)$. Given that the two shaded regions in each diagram are the same, this supports, but is not a proof of, the fact that $A - (B \cap C) = (A - B) \cup (A - C)$.

Second, two applications from logic need to be mentioned. De Morgan's Law (Theorem 1.3.2 (x)) states $\sim (P \vee Q) \equiv \sim P \wedge \sim Q$ for statements P and Q. Now let X and Y be sets such $x \notin X \cup Y$. In other words, it is not the case that $x \in X$ or $x \in Y$. By De Morgan's Law, this is equivalent to $x \notin X$ and $x \notin Y$. Thinking about it another way, note that if $x \in X$ or $x \in Y$, then $x \in X \cup Y$. Thus, given that $x \notin X \cup Y$, it must be the case that $x \notin X$ and $x \notin Y$.

Now consider the situation that $x \notin X \cap Y$. In other words, it is not the case that $x \in X$ and $x \in Y$. By De Morgan's Law (Theorem 1.3.2 (ix)), we know that $\sim (P \wedge Q) \equiv \sim P \vee \sim Q$ for statements P and Q. This means that $x \notin X \cap Y$ is equivalent to $x \notin X$ or $x \notin Y$. Thinking about it another way, note $x \notin X \cap Y$ occurs if $x \notin X$ and $x \in Y$, if $x \in X$ and $x \notin Y$, or if $x \notin X$ and $x \notin Y$. Therefore, $x \notin X \cap Y$ results in $x \notin X$ or $x \notin Y$.

Proof of Theorem 3.3.4 (vi), (vii), and (ix). To prove (vi), we start by showing $A \cap (B \cup C) \subseteq (A \cap B) \cup (A \cap C)$. Let $x \in A \cap (B \cup C)$. This means that $x \in A$ and $x \in B \cup C$. Given that $x \in B \cup C$, we know that $x \in B$ or $x \in C$. If $x \in B$, then given that $x \in A$, we have that $x \in A$ and $x \in B$ or that $x \in A \cap B$. This implies that $x \in (A \cap B) \cup (A \cap C)$. If $x \in C$, then given that $x \in A$, we have that $x \in A$ and $x \in C$ or that $x \in A \cap C$. This also implies that $x \in (A \cap B) \cup (A \cap C)$, yielding $A \cap (B \cup C) \subseteq (A \cap B) \cup (A \cap C)$.

To show $(A \cap B) \cup (A \cap C) \subseteq A \cap (B \cup C)$, let $y \in (A \cap B) \cup (A \cap C)$. This means $y \in A \cap B$ or that $y \in A \cap C$. If $y \in A \cap B$, then $y \in A$ and $y \in B$. Since $y \in B$, it follows that $y \in B \cup C$. Given that $y \in A$ and $y \in B \cup C$, we have $y \in A \cap (B \cup C)$. Now if $y \in A \cap C$, then $y \in A$ and $y \in C$. Since $y \in C$, it follows that $y \in B \cup C$. Given that $y \in A$ and $y \in B \cup C$, we have $y \in A \cap (B \cup C)$. Therefore, we see that $(A \cap B) \cup (A \cap C) \subseteq A \cap (B \cup C)$.

For (vii), we will first show $A - (B \cup C) \subseteq (A - B) \cap (A - C)$. Let $x \in A - (B \cup C)$. This means that $x \in A$ and $x \notin B \cup C$. The fact that $x \notin B \cup C$ implies that $x \notin B$ and $x \notin C$. Since $x \in A$ and $x \notin B$, we have that $x \in A - B$. But we also have $x \in A$ and $x \notin C$, implying that $x \in A - C$. The fact that $x \in A - B$ and $x \in A - C$ yields $x \in (A - B) \cap (A - C)$.

Next, to prove $(A - B) \cap (A - C) \subseteq A - (B \cup C)$, let $y \in (A - B) \cap (A - C)$. This means $y \in A - B$ and $y \in A - C$. The fact that $y \in A - B$ results in $y \in A$ and $y \notin B$. The fact that $y \in A - C$ results in $y \in A$ and $y \notin C$. Since $y \notin B$ and $y \notin C$, we know that $y \notin B \cup C$. Thus $y \in A$ and $y \notin B \cup C$ implies that $y \in A - (B \cup C)$. This gives us $(A - B) \cap (A - C) \subseteq A - (B \cup C)$.

Lastly, we will prove (ix). To show that $A \times (B \cup C) \subseteq (A \times B) \cup (A \times C)$, let $z \in A \times (B \cup C)$. This means $z = (x, y)$ where $x \in A$ and $y \in B \cup C$. The condition on y implies that $y \in B$ or $y \in C$. If $y \in B$, then $(x, y) \in A \times B$. This results in $z = (x, y) \in (A \times B) \cup (A \times C)$ or that $A \times (B \cup C) \subseteq (A \times B) \cup (A \times C)$. If $y \in C$, then $(x, y) \in A \times C$. This results in $z = (x, y) \in (A \times B) \cup (A \times C)$ or that $A \times (B \cup C) \subseteq (A \times B) \cup (A \times C)$.

To show that $(A \times B) \cup (A \times C) \subseteq A \times (B \cup C)$, let $z \in (A \times B) \cup (A \times C)$. This means $z = (x, y)$ where $(x, y) \in A \times B$ or $(x, y) \in A \times C$. If $(x, y) \in A \times B$, then $x \in A$ and $y \in B$. The fact that $y \in B$ means $y \in B \cup C$. Therefore, we have $x \in A$ and $y \in B \cup C$, indicating that $z = (x, y) \in A \times (B \cup C)$. If $(x, y) \in A \times C$, then $x \in A$ and $y \in C$. The fact that $y \in C$ means $y \in B \cup C$. Therefore, we have $x \in A$ and

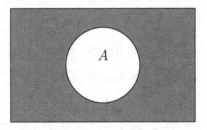

Figure 3.3 Venn diagram for $\overline{A}$.

$y \in B \cup C$, indicating that $z = (x, y) \in A \times (B \cup C)$. This shows that $(A \times B) \cup (A \times C) \subseteq A \times (B \cup C)$. □

When studying a collection of sets, you may want all of the elements in those sets to be specific types of elements. For example, when looking for possible prime numbers, you only want to examine numbers from the set $\mathbb{N}$. The set of elements considered for a certain collection of sets is called the **universe** or the **universal set**, denoted by $\mathcal{U}$. In our prime number example, you would have $\mathcal{U} = \mathbb{N}$. Now on to the final definition in this section.

Definition 3.3.5. *Let A be a set, and let $\mathcal{U}$ be the universe. The **complement** of set A is the set $\overline{A} = \mathcal{U} - A$.*

Simply put, the complement of a set A is all the elements of $\mathcal{U}$ not in A. The Venn diagram for the complement of a set A is given in Figure 3.3 with the shaded region being $\overline{A}$.

Example 3.3.6. *Consider the sets $A = \{0, 2, 4, 6, 8\}, B = \{4, 5, 6, 7\}$, and $\mathcal{U} = \{0, 1, 2, 3, 4, 5, 6, 7, 8, 9\}$.*

- $\overline{A} = \{1, 3, 5, 7, 9\}$

- $\overline{A \cup B} = \overline{\{0, 2, 4, 5, 6, 7, 8\}} = \{1, 3, 9\}$

- $A \cap \overline{B} = \{0, 2, 4, 6, 8\} \cap \{0, 1, 2, 3, 8, 9\} = \{0, 2, 8\}$

- $\overline{A - B} = \overline{\{1, 3, 9\}} = \{0, 2, 4, 5, 6, 7, 8\}$

- $\overline{\emptyset} = \mathcal{U}$

Theorem 3.3.7. *Let A and B be sets, and let $\mathcal{U}$ be the universe.*

   (i) $\overline{\overline{A}} = A$

   (ii) $\overline{A \cup B} = \overline{A} \cap \overline{B}$ *De Morgan's Law*

   (iii) $\overline{A \cap B} = \overline{A} \cup \overline{B}$ *De Morgan's Law*

Proof of Theorem 3.3.7 (iii). To prove $\overline{A \cap B} \subseteq \overline{A} \cup \overline{B}$, let $x \in \overline{A \cap B}$. This means $x \notin A \cap B$. This occurs when $x \notin A$ or $x \notin B$. If $x \notin A$, then it follows that $x \in \overline{A}$. We then have $x \in \overline{A} \cup \overline{B}$. If $x \notin B$, then it follows that $x \in \overline{B}$. We then have $x \in \overline{A} \cup \overline{B}$. This yields $\overline{A \cap B} \subseteq \overline{A} \cup \overline{B}$.

    To show that $\overline{A} \cup \overline{B} \subseteq \overline{A \cap B}$, let $y \in \overline{A} \cup \overline{B}$. This implies that $y \in \overline{A}$ or that $y \in \overline{B}$. If $y \in \overline{A}$, then $y \notin A$. This tells us that $y \notin A \cap B$, which results in $y \in \overline{A \cap B}$. If $y \in \overline{B}$, then $y \notin B$. This again tells us that $y \notin A \cap B$, which results in $y \in \overline{A \cap B}$. Consequently, we have $\overline{A} \cup \overline{B} \subseteq \overline{A \cap B}$.   □

    One final set theoretic concept needs to be mentioned before this section ends. It will be needed in a few of the exercises here and in Section 3.4. Let X be a set and x be any element from the universe. Applying Theorem 1.3.1 (i), which states for a statement P that $P \vee \sim P$ is a tautology, it follows that $x \in X$ or that $x \notin X$.

Exercises 3.3

(1) Let $A = \{3, 5, 7, 11, 13, 17, 19\}, B = \{2, 11, 23, 31\}, C = \{3, 17, 29\}$, and $D = \{17, 19, 23, 29, 31\}$ with universe $\mathcal{U} = \{2, 3, 5, 7, 11, 13, 17, 19, 23, 29, 31\}$. Determine the following sets.

    (a) $A \cap B$ (f) $A \cap C \cap D$

    (b) $(A \cup B) - C$ (g) $\overline{A \cup B}$

    (c) $D - (B \cap C)$ (h) $\overline{\overline{A} \cup C}$

    (d) $(B \cap C) - D$ (i) $(A - B) \cup (B - A)$

    (e) $\overline{B}$ (j) $A - \overline{C \cup D}$

(2) Given an example of sets A and B such that $A \cup B \subseteq A \cap B$.

(3) Given an example of sets A and B such that $A \neq B$ and $A - B \subseteq B - A$.

(4) Given sets A, B and C, draw two Venn diagrams to provide evidence that $A - (B \cup C) = (A - B) \cap (A - C)$.

(5) Given sets A and B, draw two Venn Diagrams to provide evidence that $\overline{A \cap B} = \overline{A} \cup \overline{B}$.

(6) Prove Lemma 3.3.3 (ii): for A a set, $A \cup \emptyset = A$.

(7) Prove Lemma 3.3.3 (iii): for A a set, $A \cap A = A$.

(8) Prove Theorem 3.3.4 (ii): for A and B sets, $A \cap B = B \cap A$.

(9) Prove Theorem 3.3.4 (iv): for A, B and C sets, $A \cap (B \cap C) = (A \cap B) \cap C$.

(10) Prove Theorem 3.3.4 (v): for A, B, and C sets, $A \cup (B \cap C) = (A \cup B) \cap (A \cup C)$.

(11) Prove Theorem 3.3.4 (viii): for A, B and C sets, $A - (B \cap C) = (A - B) \cup (A - C)$.

(12) Prove Theorem 3.3.4 (x): for sets A, B and C, $A \times (B \cap C) = (A \times B) \cap (A \times C)$.

(13) Given sets A and B, prove that $(A - B) \cap (B - A) = \emptyset$.

(14) Let A and B be sets such that $B \subseteq A$. Prove that $(A - B) \cup B = A$.

(15) Show that $(A - B) \cup B = A$ is not necessarily true for arbitrary sets A and B.

(16) Given sets A, B, and C such that $A \cap C = B \cap C = \emptyset$, prove that $A \cup C = B \cup C$ if and only if $A = B$.

(17) Given sets A and B, prove that $A \cap B \subseteq A \subseteq A \cup B$.

(18) Given sets A and B, prove that $A \cup (A \cap B) = A$.

(19) Given sets A and B, prove that $A \cup (B - A) = A \cup B$.

(20) Given sets A and B, prove that $(A \cup B) - (A \cap B) = (A - B) \cup (B - A)$.

(21) Given sets A, B, and C, prove that $(A \cup B) - C = (A - C) \cup (B - C)$.

(22) Given sets A, B, and C, prove that $(A \cap B) - C = (A - C) \cap (B - C)$.

(23) Given sets A, B, and C, prove that if $A \subseteq B$, then $A - C \subseteq B - C$.

(24) Given sets A and B, prove that $A \cup B = A$ if and only if $B \subseteq A$.

(25) Given sets A and B, prove that $A \cap B = A$ if and only if $A \subseteq B$.

(26) Given sets A, B, C, and D, prove that $(A \times B) \cap (C \times D) = (A \cap C) \times (B \cap D)$.

(27) Given sets A, B, C, and D, prove that $(A \times B) \cup (C \times D) \subseteq (A \cup C) \times (B \cup D)$.

(28) Given sets A and B, prove that $\mathcal{P}(A) \cap \mathcal{P}(B) = \mathcal{P}(A \cap B)$.

(29) Prove Theorem 3.3.7 (i): for A a set, $\overline{\overline{A}} = A$.

(30) Prove Theorem 3.3.7 (ii): for A and B sets, $\overline{A \cup B} = \overline{A} \cap \overline{B}$.

(31) Given sets A and B, prove that $A - B = A \cap \overline{B}$.

3.4 INDEXED SETS

There may be situations when using the letters $A, B, C, D, \ldots$ is too cumbersome or too limiting to list a collection of sets in which you are interested. When this occurs, subscripts can be used to create **indexed sets**. The collection of subscripts is called the **index set**. For example, consider the following collection of indexed sets.

$$B_1 = \{1\}, B_2 = \{1, 2\}, B_3 = \{1, 2, 3\}, \ldots, B_{50} = \{1, 2, 3, \ldots, 50\}$$

Given that the integers $1, 2, 3, \ldots, 50$ are used as subscripts, the index set is $\{1, 2, 3, \ldots, 50\}$.

In general, the set $\{1, 2, 3, \ldots, n\}$, for n a positive integer, is a common index set. When given a collection of sets $A_1, A_2, A_3, \ldots, A_n$, indexed by $\{1, 2, 3, \ldots, n\}$, the following two natural computations can be defined.

$$\bigcup_{i=1}^{n} A_i = A_1 \cup A_2 \cup A_3 \cup \cdots \cup A_n = \{a : a \in A_i \text{ for some } i \in \{1, 2, 3, \ldots, n\}\}$$

$$\bigcap_{i=1}^{n} A_i = A_1 \cap A_2 \cap A_3 \cap \cdots \cap A_n = \{a : a \in A_i \text{ for every } i \in \{1, 2, 3, \ldots, n\}\}$$

From the example above, we see that

$$\bigcup_{i=1}^{50} B_i = B_1 \cup B_2 \cup B_3 \cup \cdots \cup B_{50}$$

$$= \{1\} \cup \{1,2\} \cup \{1,2,3\} \cup \cdots \cup \{1,2,3,\ldots,50\}$$
$$= \{1,2,3,\ldots,50\}$$
$$= B_{50}$$

and

$$\bigcap_{i=1}^{50} B_i = B_1 \cap B_2 \cap B_3 \cap \cdots \cap B_{50}$$

$$= \{1\} \cap \{1,2\} \cap \{1,2,3\} \cap \cdots \cap \{1,2,3,\ldots,50\}$$
$$= \{1\}$$
$$= B_1.$$

For another example, consider the collection of sets $C_i = [i, i+1]$, for $i \in \{1,2,3,\ldots,20\}$. For this collection, we have

$$\bigcup_{i=1}^{20} C_i = C_1 \cup C_2 \cup C_3 \cup \cdots \cup C_{20}$$

$$= [1,2] \cup [2,3] \cup [3,4] \cup \cdots \cup [20,21]$$
$$= [1,21]$$

and

$$\bigcap_{i=1}^{20} C_i = C_1 \cap C_2 \cap C_3 \cap \cdots \cap C_{20}$$

$$= [1,2] \cap [2,3] \cap [3,4] \cap \cdots \cap [20,21]$$
$$= \emptyset.$$

You can have an infinite collection of indexed sets. For example, consider the collection of sets $D_i = \{-i, 0, i\}$ for $i \in \mathbb{N}$. In other words, we have $D_1 = \{-1, 0, 1\}, D_2 = \{-2, 0, 2\}, D_3 = \{-3, 0, 3\}$, and so on. Here the index set is the natural numbers $\mathbb{N}$. This motivates the following computations. Given a collection of sets A_i with index set $\mathbb{N}$, then

$$\bigcup_{i=1}^{\infty} A_i = A_1 \cup A_2 \cup A_3 \cup A_4 \cup \cdots = \{a : a \in A_i \text{ for some } i \in \mathbb{N}\}$$

and

$$\bigcap_{i=1}^{\infty} A_i = A_1 \cap A_2 \cap A_3 \cap A_4 \cap \cdots = \{a : a \in A_i \text{ for every } i \in \mathbb{N}\}.$$

For the collection of sets D_i, with $i \in \mathbb{N}$, defined above, we have

$$\bigcup_{i=1}^{\infty} D_i = D_1 \cup D_2 \cup D_3 \cup D_4 \cup \cdots$$

$$= \{-1, 0, 1\} \cup \{-2, 0, 2\} \cup \{-3, 0, 3\} \cup \{-4, 0, 4\} \cup \cdots$$
$$= \{\cdots, -4, -3, -2, -1, 0, 1, 2, 3, 4, \ldots\}$$
$$= \mathbb{Z}$$

and

$$\bigcap_{i=1}^{\infty} D_i = D_1 \cap D_2 \cap D_3 \cap D_4 \cup \cdots$$

$$= \{-1, 0, 1\} \cap \{-2, 0, 2\} \cap \{-3, 0, 3\} \cap \{-4, 0, 4\} \cap \cdots$$
$$= \{0\}.$$

For another example, consider the collection of sets $E_n = [0, 1 + \frac{1}{n})$, for $n \in \mathbb{N}$. For this collection,

$$\bigcup_{n=1}^{\infty} E_n = E_1 \cup E_2 \cup E_3 \cup E_4 \cup \cdots$$

$$= \left[0, 1 + \frac{1}{1}\right) \cup \left[0, 1 + \frac{1}{2}\right) \cup \left[0, 1 + \frac{1}{3}\right) \cup \left[0, 1 + \frac{1}{4}\right) \cup \cdots$$
$$= [0, 2) \cup \left[0, \frac{3}{2}\right) \cup \left[0, \frac{4}{3}\right) \cup \left[0, \frac{5}{4}\right) \cup \cdots$$
$$= [0, 2)$$
$$= E_1$$

and

$$\bigcap_{n=1}^{\infty} E_n = E_1 \cap E_2 \cap E_3 \cap E_4 \cap \cdots$$

$$= \left[0, 1 + \frac{1}{1}\right) \cap \left[0, 1 + \frac{1}{2}\right) \cap \left[0, 1 + \frac{1}{3}\right) \cap \left[0, 1 + \frac{1}{4}\right) \cap \cdots$$
$$= [0, 2) \cap \left[0, \frac{3}{2}\right) \cap \left[0, \frac{4}{3}\right) \cap \left[0, \frac{5}{4}\right) \cap \cdots$$
$$= [0, 1]$$

To see that $\bigcup_{n=1}^{\infty} E_n = [0, 2) = E_1$, note that for $n \in \mathbb{N}$, with $n \geq 2$, that $\frac{1}{n} < 1$. This implies that $1 + \frac{1}{n} < 1 + 1 = 2$ or that $E_n \subseteq E_1 = [0, 2)$. Since $E_n \subseteq E_1$, for $n \geq 2$, it follows that $\bigcup_{i=1}^{\infty} E_i = [0, 2) = E_1$.

To see that $\bigcap_{n=1}^{\infty} E_n = [0, 1]$, first note for $n \in \mathbb{N}$, that $\frac{1}{n} > 0$. This implies that $1 + \frac{1}{n} > 1$ for all n. Thus for every n, it follows that $[0, 1] \subseteq [0, 1 + \frac{1}{n})$. Thus at the very least we know $[0, 1] \subseteq \bigcap_{n=1}^{\infty} E_n$. Now suppose there is a real number y, such that $y > 1$ and $y \in \bigcap_{n=1}^{\infty} E_n$. This means for every $n \in \mathbb{N}$ that $y \in E_n = [0, 1 + \frac{1}{n})$. However, we know from calculus that $\lim_{n \to \infty} 1 + \frac{1}{n} = 1$. This means given that $y > 1$, there exists an $N \in \mathbb{N}$ such that $1 + \frac{1}{N} < y$. Thus y is not in the set $E_N = [0, 1 + \frac{1}{N})$. This is a contradiction as y must be in E_n, for all $n \in \mathbb{N}$, to be in the intersection. This results in $\bigcap_{n=1}^{\infty} E_n = [0, 1]$.

Any set I can be an index set for a collection of sets. The collection of sets $F_r = [-|r|, 0]$ for $r \in \mathbb{Q}$ has the rational numbers $\mathbb{Q}$ as its index set. The collection of sets $G_t = \{t^2\}$ for $t \in \mathbb{R}$ is indexed by the set of real numbers $\mathbb{R}$. We can now make the following definitions that generalize the ideas previously presented.

Definition 3.4.1. *Given a collection of sets A_i indexed by a set I, then*

$$\bigcup_{i \in I} A_i = \{a : a \in A_i \text{ for some } i \in I\}$$

and

$$\bigcap_{i \in I} A_i = \{a : a \in A_i \text{ for every } i \in I\}.$$

From our examples above, we get the following.

$$\bigcup_{r \in \mathbb{Q}} F_r = \bigcup_{r \in \mathbb{Q}} [-|r|, 0] \qquad \bigcap_{r \in \mathbb{Q}} F_r = \bigcap_{r \in \mathbb{Q}} [-|r|, 0]$$
$$= \bigcup_{r \in \mathbb{Q}} \{x \in \mathbb{R} : -|r| \leq x \leq 0\} \qquad = \bigcap_{r \in \mathbb{Q}} \{x \in \mathbb{R} : -|r| \leq x \leq 0\}$$
$$= (-\infty, 0] \qquad = \{0\}$$

$$\bigcup_{t \in \mathbb{R}} G_t = \bigcup_{t \in \mathbb{R}} \{t^2\} \qquad \bigcap_{t \in \mathbb{R}} G_t = \bigcap_{t \in \mathbb{R}} \{t^2\}$$
$$= \{y \in \mathbb{R} : y > 0\} \qquad = \emptyset$$
$$= [0, \infty)$$

Given an indexed collection of sets, you can generalize properties proven earlier concerning intersection, union, difference, and complement.

Theorem 3.4.2. *Let A_i be an indexed collection of sets indexed by set I, and let B be a set.*

(i) $\quad B \cup \left(\bigcap_{i \in I} A_i \right) = \bigcap_{i \in I} (B \cup A_i) \qquad$ *Distributive Law*

(ii) $\quad B \cap \left(\bigcup_{i \in I} A_i \right) = \bigcup_{i \in I} (B \cap A_i) \qquad$ *Distributive Law*

(iii) $\quad \overline{\bigcup_{i \in I} A_i} = \bigcap_{i \in I} \overline{A_i} \qquad$ *De Morgan's Law*

(iv) $\quad \overline{\bigcap_{i \in I} A_i} = \bigcup_{i \in I} \overline{A_i} \qquad$ *De Morgan's Law*

(v) $\quad B - \left(\bigcup_{i \in I} A_i \right) = \bigcap_{i \in I} (B - A_i) \qquad$ *De Morgan's Law*

(vi) $\quad B - \left(\bigcap_{i \in I} A_i \right) = \bigcup_{i \in I} (B - A_i) \qquad$ *De Morgan's Law*

Only items (i), (iii), and (vi) from Theorem 3.4.2 will be proven here and the rest are exercises.

Proof of Theorem 3.4.2 (i), (iii), and (vi). To prove (i), we start by showing $B \cup (\bigcap_{i \in I} A_i) \subseteq \bigcap_{i \in I} (B \cup A_i)$. Let $x \in B \cup (\bigcap_{i \in I} A_i)$. This means $x \in B$ or $x \in \bigcap_{i \in I} A_i$. If $x \in B$, then $x \in B \cup A_i$ for all $i \in I$. This implies that $x \in \bigcap_{i \in I} (B \cup A_i)$. If $x \in \bigcap_{i \in I} A_i$, then $x \in A_i$ for all $i \in I$. This means $x \in B \cup A_i$ for all i, which yields $x \in \bigcap_{i \in I} (B \cup A_i)$.

To show that $\bigcap_{i \in I} (B \cup A_i) \subseteq B \cup (\bigcap_{i \in I} A_i)$, let $y \in \bigcap_{i \in I} (B \cup A_i)$. This results in $y \in B \cup A_i$ for all $i \in I$. If $y \in B$, then $y \in B \cup (\bigcap_{i \in I} A_i)$. Suppose that $y \notin B$. Since $y \in B \cup A_i$ for all i, this implies that $y \in A_i$ for all i. Then we have $y \in \bigcap_{i \in I} A_i$, resulting in $y \in B \cup (\bigcap_{i \in I} A_i)$. It then follows that $\bigcap_{i \in I} (B \cup A_i) \subseteq B \cup (\bigcap_{i \in I} A_i)$.

For (iii), we begin by showing $\overline{\bigcup_{i \in I} A_i} \subseteq \bigcap_{i \in I} \overline{A_i}$. Let $x \in \overline{\bigcup_{i \in I} A_i}$. This means $x \notin \bigcup_{i \in I} A_i$. Thus for all $i \in I$, it must be that $x \notin A_i$. Then we have $x \in \overline{A_i}$ for all $i \in I$, which yields $x \in \bigcap_{i \in I} \overline{A_i}$. This results in $\overline{\bigcup_{i \in I} A_i} \subseteq \bigcap_{i \in I} \overline{A_i}$.

To show that $\bigcap_{i \in I} \overline{A_i} \subseteq \overline{\bigcup_{i \in I} A_i}$, let $y \in \bigcap_{i \in I} \overline{A_i}$. This means that $y \in \overline{A_i}$ for all $i \in I$. Therefore, we have that $y \notin A_i$ for all i. Given this fact, we now know that $y \notin \bigcup_{i \in I} A_i$ or that $y \in \overline{\bigcup_{i \in I} A_i}$. This shows that $\bigcap_{i \in I} \overline{A_i} \subseteq \overline{\bigcup_{i \in I} A_i}$.

Lastly, to prove (vi) is true, we begin by showing $B - (\bigcap_{i \in I} A_i) \subseteq \bigcup_{i \in I} (B - A_i)$. Let $x \in B - (\bigcap_{i \in I} A_i)$. This means $x \in B$ and that

$x \notin \bigcap_{i \in I} A_i$. Given that $x \notin \bigcap_{i \in I} A_i$, it follows that there exists a $j \in I$ such that $x \notin A_j$. Since $x \in B$ and $x \notin A_j$, it follows that $x \in B - A_j$. This results in $x \in \bigcup_{i \in I} (B - A_i)$, which shows $B - (\bigcap_{i \in I} A_i) \subseteq \bigcup_{i \in I} (B - A_i)$.

To prove that $\bigcup_{i \in I} (B - A_i) \subseteq B - (\bigcap_{i \in I} A_i)$, let $y \in \bigcup_{i \in I} (B - A_i)$. Thus for some $k \in I$, we know that $y \in B - A_k$. This implies that $y \in B$ and $y \notin A_k$. Given that $y \notin A_k$, we know that $y \notin \bigcap_{i \in I} A_i$. Given that $y \in B$ and $y \notin \bigcap_{i \in I} A_i$, it follows that $y \in B - (\bigcap_{i \in I} A_i)$. We have now shown that $\bigcup_{i \in I} (B - A_i) \subseteq B - (\bigcap_{i \in I} A_i)$. $\qquad\square$

Exercises 3.4

(1) For the sets $A_i = \{0, 2i\}$, with $i \in \{1, 2, \ldots, 30\}$, determine the sets $\bigcup_{i=1}^{30} A_i$ and $\bigcap_{i=1}^{30} A_i$.

(2) For the sets $A_i = \{\frac{-1}{i}, \frac{1}{i}\}$, with $i \in \{1, 2, \ldots, 10\}$, determine the sets $\bigcup_{i=1}^{10} A_i$ and $\bigcap_{i=1}^{10} A_i$.

(3) For the sets $A_i = [\frac{-1}{i}, \frac{1}{i}]$, with $i \in \{1, 2, \ldots, 100\}$, determine the sets $\bigcup_{i=1}^{100} A_i$ and $\bigcap_{i=1}^{100} A_i$.

(4) For the sets $B_n = \{-2n, 0, 2n\}$, with $n \in \mathbb{N}$, determine the sets $\bigcup_{n=1}^{\infty} B_n$ and $\bigcap_{n=1}^{\infty} B_n$.

(5) For the sets $B_n = \{\frac{1}{n}\}$, with $n \in \mathbb{N}$, determine the sets $\bigcup_{n=1}^{\infty} B_n$ and $\bigcap_{n=1}^{\infty} B_n$.

(6) For the sets $B_n = [\frac{-1}{n}, 1]$, with $n \in \mathbb{N}$, determine the sets $\bigcup_{n=1}^{\infty} B_n$ and $\bigcap_{n=1}^{\infty} B_n$.

(7) For the sets $B_n = [-1, 1 - \frac{1}{n}]$, with $n \in \mathbb{N}$, determine the sets $\bigcup_{n=1}^{\infty} B_n$ and $\bigcap_{n=1}^{\infty} B_n$.

(8) For the sets $B_n = [-n, \frac{1}{n}]$, with $n \in \mathbb{N}$, determine the sets $\bigcup_{n=1}^{\infty} B_n$ and $\bigcap_{n=1}^{\infty} B_n$.

(9) For the sets $C_r = [|r|, \infty)$, with $r \in \mathbb{Q}$, determine the sets $\bigcup_{r \in \mathbb{Q}} C_r$ and $\bigcap_{r \in \mathbb{Q}} C_r$.

(10) For the sets $C_r = \{|t|\}$, with $t \in \mathbb{R}$, determine the sets $\bigcup_{t \in \mathbb{R}} C_t$ and $\bigcap_{t \in \mathbb{R}} C_t$.

(11) Determine a collection of sets A_n, with $n \in \mathbb{N}$, such that $\bigcup_{n=1}^{\infty} A_n = [0, 2)$ and $\bigcap_{n=1}^{\infty} A_n = \{0\}$.

(12) Determine a collection of sets A_n, with $n \in \mathbb{N}$, such that $\bigcup_{n=1}^{\infty} A_n = \bigcap_{n=1}^{\infty} A_n$.

(13) For a collection of sets A_i indexed by set I, prove that $\bigcap_{i \in I} A_i \subseteq \bigcup_{i \in I} A_i$.

(14) For a set B and collection of sets A_i indexed by set I, such that $B \subseteq A_i$ for all $i \in I$, prove that $B \subseteq \bigcap_{i \in I} A_i$.

(15) Prove Theorem 3.4.2 (ii): for A_i an indexed collection of sets indexed by set I, and a set B, then $B \cap (\bigcup_{i \in I} A_i) = \bigcup_{i \in I} (B \cap A_i)$.

(16) Prove Theorem 3.4.2 (iv): for A_i an indexed collection of sets indexed by set I, then $\overline{\bigcap_{i \in I} A_i} = \bigcup_{i \in I} \overline{A_i}$.

(17) Prove Theorem 3.4.2 (v): for A_i an indexed collection of sets indexed by set I, and a set B, then $B - (\bigcup_{i \in I} A_i) = \bigcap_{i \in I} (B - A_i)$.

(18) Given a collection of sets A_i, indexed by set I, and a set B, prove that $(\bigcup_{i \in I} A_i) - B = \bigcup_{i \in I} (A_i - B)$.

(19) Given a collection of sets A_i, indexed by set I, and a set B, prove that $(\bigcap_{i \in I} A_i) - B = \bigcap_{i \in I} (A_i - B)$.

(20) Given a collection of sets B_i, indexed by set I, and a set A, prove that $A \times (\bigcup_{i \in I} B_i) = \bigcup_{i \in I} (A \times B_i)$.

(21) Given a collection of sets B_i, indexed by set I, and a set A, prove that $A \times (\bigcap_{i \in I} B_i) = \bigcap_{i \in I} (A \times B_i)$.

3.5   RUSSELL'S PARADOX

We began this chapter by agreeing on the notion of a set. We did not give a formal definition. This began our journey down the road of naive set theory. One of the motivations for this is an observation by the philosopher and mathematician Bertrand Russell (1872-1970) called **Russell's Paradox**. It should be noted that this paradox was also independently discovered by the logician and mathematician Ernst Zermelo (1871-1953).

Consider the set S defined by

$$S = \{X : X \text{ is a set such that } X \notin X\}.$$

It is the set of all sets that do not have themselves as elements. The set

S is nonempty as there are numerous sets in it. For example, the set $\{1, 2, 3\}$ is in S, given that $\{1, 2, 3\} \notin \{1, 2, 3\}$, and the set $\mathbb{Q} \in S$ since $\mathbb{Q} \notin \mathbb{Q}$ (the set of rational numbers is not itself a rational number).

However, there are sets that are not in S. Consider the set Y defined by

$$Y = \{\{\{\{\cdots\}\}\}\} = \{\{\{\underbrace{\{\cdots\}}_{Y}\}\}\}.$$

Given that $Y \in Y$, it follows that Y is an element of itself and that $Y \notin S$. Russell's Paradox comes from determining if $S \in S$. There are two cases to examine.

Case (1) $S \in S$ is true. Given that the only elements of set S are sets that do not have themselves as an element, then $S \in S$ implies that $S \notin S$. This results in $S \in S$ being a false statement. This is a contradiction, as a statement can't be both true and false.

Case (2) $S \in S$ is false. This means that $S \notin S$. Since the only elements of set S are sets that do not have themselves as elements, the fact that $S \notin S$ implies that $S \in S$ is true. This is again a contradiction, as a statement can't be both true and false.

One way to think of this paradox is to consider the following statement and question. In the town of Aurora, barber Joe only shaves individuals who do not shave themselves. Who shaves the barber? If the barber shaves himself, then by the statement, the barber does not shave himself. This is a contradiction. If the barber does shave himself, then by the statement, the barber does not shave himself, which is another contradiction.

Russell's Paradox and other paradoxes (we will discuss Cantor's Paradox in Chapter 7) came to light in the late nineteenth and early twentieth centuries. This led to a reexamination of the foundations of mathematics. To get set theory back on a solid footing that avoided these paradoxes, Ernst Zermelo established a list of axioms for set theory in 1908. These were slightly modified, upon suggestions by Abraham Fraenkel (1891-1965) and others, to create a set of axioms for set theory that are referred to as the Zermelo-Fraenkel axioms.

The term naive set theory comes from working with sets using a minimal number or no axioms. When doing so, you are not concerned about Russell's Paradox. One approach to naive set theory can be found in the book [Hal17]. If you are interested in learning more about Russell's

Paradox, the Zermelo-Fraenkel axioms for set theory, and a more general introduction to axiomatic set theory, I suggest the books [Sup72] and [End77].

CHAPTER 4

Proof by Mathematical Induction

4.1 INTRODUCTION

Statements in mathematics of the form $(\forall n \in \mathbb{N}), P(n)$, for predicate $P(n)$, occur often in mathematics. The following statements are two examples.

- For all $n \in \mathbb{N}, 1 + 2 + 3 + \cdots + n = \frac{n(n+1)}{2}$.

- For all $n \in \mathbb{N}, 4|(5^n - 1)$.

Given that these statements are said to be true for every natural number, we need to take a quick look at the natural numbers $\mathbb{N} = \{1, 2, 3, \ldots\}$ before we discuss how to prove these types of statements.

To begin, note that there are sets with a least or smallest element and there are sets with no least or smallest element. The finite set $\{2, 4, 6, 8\}$ has 2 as its least element as every element in the set is greater than or equal to 2. In fact, every finite set of numbers has a least element. But what about infinite sets? The infinite set $\{10, 20, 30, \ldots\}$ has 10 as a least element, but the infinite set $\{\ldots, -3, -2, -1\}$ has no least element. In addition, the interval $(0, 1)$ does not have a least element. To show this, suppose the interval $(0, 1)$ has a least element x. It follows that $0 < x < 1$ and that $0 < \frac{x}{2} < \frac{1}{2}$. Thus the real number $\frac{x}{2} \in (0, 1)$ with $\frac{x}{2} < x$. This is a contradiction as we assumed that x was the least element in $(0, 1)$. Therefore, the interval $(0, 1)$ has no smallest element.

In the examples from the previous paragraph, the sets that were subsets of $\mathbb{N}$ had smallest elements. This motivates the following fundamental axiom concerning the natural numbers.

Axiom 4.1.1 (Well-Ordering Principle). *Every non-empty subset of the natural numbers $\mathbb{N}$ has a least element.*

Axiom 4.1.1 has numerous applications. To start, it is used in the proof of the Division Algorithm (Theorem 2.3.8). If you are interested in this proof, it can be found in [JJ98].

By the Well-Ordering Principle, the natural numbers $\mathbb{N}$ have a least (smallest) element 1 with every number in $\mathbb{N}$ greater than or equal to 1. Using the Well-Ordering Principle again, we see that the set $\mathbb{N} - \{1\}$ has a least element 2. If you examine the set $\mathbb{N} - \{1, 2\}$, it has a least element 3. Continuing in this way, given any natural number n, then $n + 1$ is a natural number and the least element of the set $\mathbb{N} - \{1, 2, \ldots, n\}$. This results in the natural ordering $1, 2, 3, 4, \ldots$ of the numbers in $\mathbb{N}$.

Now consider a non-empty subset S of $\mathbb{N}$ such that $1 \in S$ and $s + 1 \in S$ when $s \in S$. Suppose S does not equal $\mathbb{N}$. Then $\mathbb{N} - S$ is non-empty and a subset of $\mathbb{N}$. By the Well-Ordering Principle (Axiom 4.1.1), $\mathbb{N} - S$ has a least element n. Since $1 \in S$, it follows that $1 \notin \mathbb{N} - S$ and $n \geq 2$. Consider the natural number $n - 1$. Since $n - 1 < n$ and n is the least element if $\mathbb{N} - S$, we have $n - 1 \notin \mathbb{N} - S$ and $n - 1 \in S$. But this implies that $(n - 1) + 1 = n$ is in S. This contradiction means $S = \mathbb{N}$. This is summed up in the following lemma, which, as we have just shown, follows directly from the Well-Ordering Principle.

Lemma 4.1.2. *Let S be a subset of the natural numbers. If $1 \in S$ and $s \in S$ implies $s + 1$ is in S, then $S = \mathbb{N}$.*

4.2 THE PRINCIPLE OF MATHEMATICAL INDUCTION

Now let's return to statements of the form $(\forall n \in \mathbb{N}), P(n)$. Lemma 4.1.2 provides the basis of the proof technique.

Theorem 4.2.1 (Principle of Mathematical Induction-Version 1). *Let $P(n)$ be a statement for every $n \in \mathbb{N}$. If both*

(i) *$P(1)$ is true and*

(ii) *the implication $P(k)$ implies $P(k + 1)$ is true for all $k \in \mathbb{N}$,*

then $P(n)$ is true for all $n \in \mathbb{N}$.

Proving a statement using the Principle of Mathematical Induction is commonly referred to as **proof by mathematical induction** or more simply as **proof by induction**. An easy way to think about it is the task of climbing a ladder that has infinitely many rungs and you want to climb to the nth rung of the ladder for some rung $n \in \mathbb{N}$. The statement, "$P(1)$ is true" signifies that you can climb to the first rung of the ladder. The statement, "the implication $P(k)$ implies $P(k+1)$ is true for all $k \in \mathbb{N}$" means that if you are on rung k, you can always climb to the next rung $k+1$. Since you have climbed to the first rung or rung 1, you are on the ladder. Now that you are on the first rung or rung 1, you can climb to rung $1+1$ or the second rung of the ladder. But now that you are on the second rung or rung 2, you can climb to rung $2+1$ or the third rung of the ladder. In this way, you repeat the process until you have climbed to the desired n^{th} rung.

When using proof by induction to prove a statement of the form $(\forall n \in \mathbb{N}), P(n)$, you start by showing that $P(1)$ is true. This is called the **base case** and should be appropriately labeled in your proof. Next, you show that $P(k)$ implies $P(k+1)$ is true for $k \in \mathbb{N}$. This is called the **inductive case** and should also be labeled as such in your proof. The assumption that $P(k)$ is true, which is used in the inductive case, is called the **induction hypothesis**. Now let's dig in.

Proposition 4.2.2. *For all $n \in \mathbb{N}$, $1 + 2 + 3 + \cdots + n = \frac{n(n+1)}{2}$.*

Proof. Let $P(n)$ denote the statement $1 + 2 + 3 + \cdots + n = \frac{n(n+1)}{2}$.

Base Case: We must show that $P(1)$ is true. When $n = 1$, the sum is 1 and $\frac{n(n+1)}{2} = \frac{1(1+1)}{2} = \frac{2}{2} = 1$. Since the resulting values are equal, $P(1)$ is true.

Inductive Case: Let $k \in \mathbb{N}$ and assume that $P(k)$ is true. Thus we are assuming that

$$1 + 2 + 3 + \cdots + k = \frac{k(k+1)}{2}.$$

We must show $P(k+1)$ is true or that

$$1 + 2 + 3 + \cdots + k + (k+1) = \frac{(k+1)((k+1)+1)}{2} = \frac{(k+1)(k+2)}{2}.$$

The calculation

$$1 + 2 + 3 + \cdots + k + (k+1) = (1 + 2 + 3 + \cdots + k) + (k+1)$$
$$= \frac{k(k+1)}{2} + k + 1$$
$$= \frac{k(k+1)}{2} + \frac{2(k+1)}{2}$$

$$= \frac{k(k+1) + 2(k+1)}{2}$$

$$= \frac{(k+1)(k+2)}{2}$$

shows that the implication $P(k)$ implies $P(k+1)$ is true. □

Proposition 4.2.3. *For all $n \in \mathbb{N}$, $1^2 + 2^2 + 3^2 + \cdots + n^2 = \frac{n(n+1)(2n+1)}{6}$.*

Proof. Let $P(n)$ denote the statement $1^2 + 2^2 + 3^2 + \cdots + n^2 = \frac{n(n+1)(2n+1)}{6}$.

Base Case: We must show that $P(1)$ is true. When $n = 1$, the sum is $1^2 = 1$ and $\frac{n(n+1)(2n+1)}{6} = \frac{1(1+1)(2\cdot1+1)}{6} = \frac{1\cdot2\cdot3}{6} = \frac{6}{6} = 1$. Since the resulting values are equal, $P(1)$ is true.

Inductive Case: Let $k \in \mathbb{N}$ and assume that $P(k)$ is true. Thus we are assuming that

$$1^2 + 2^2 + 3^2 + \cdots + k^2 = \frac{k(k+1)(2k+1)}{6}.$$

We must show $P(k+1)$ is true or that

$$1^2 + 2^2 + 3^2 + \cdots + k^2 + (k+1)^2 = \frac{(k+1)((k+1)+1)(2(k+1)+1)}{6}$$

$$= \frac{(k+1)(k+2)(2k+3)}{6}.$$

The calculation

$$1^2 + 2^2 + 3^2 + \cdots + k^2 + (k+1)^2 = (1^2 + 2^2 + 3^2 + \cdots + k^2) + (k+1)^2$$

$$= \frac{k(k+1)(2k+1)}{6} + (k+1)^2$$

$$= \frac{k(k+1)(2k+1) + 6(k+1)^2}{6}$$

$$= \frac{(k+1)[k(2k+1) + 6(k+1)]}{6}$$

$$= \frac{(k+1)[2k^2 + 7k + 6]}{6}$$

$$= \frac{(k+1)(k+2)(2k+3)}{6}$$

shows that the implication $P(k)$ implies $P(k+1)$ is true. □

Proposition 4.2.4. *For all $n \in \mathbb{N}$, $4|(5^n - 1)$.*

Proof. Let $P(n)$ denote the statement $4|(5^n - 1)$.

Base Case: We must show that $P(1)$ is true. When $n = 1$, it follows that $5^n - 1 = 5^1 - 1 = 5 - 1 = 4 = 1 \cdot 4$. Since 1 is an integer, $4|(5^n - 1)$ when $n = 1$ and we see that statement $P(1)$ is true.

Inductive Case: Let $k \in \mathbb{N}$ and assume that $P(k)$ is true. This means

$$4|(5^k - 1),$$

implying that there exists an integer l such that $5^k - 1 = 4l$ or $5^k = 4l+1$. We must show that $P(k+1)$ is true or that

$$4|(5^{k+1} - 1).$$

Rewriting, we get that $5^{k+1} - 1 = 5^k 5^1 - 1 = 5^k 5 - 1$. Given that $5^k = 4l + 1$, we can substitute it into the previous equation to get $5^{k+1} - 1 = (4l + 1)5 - 1$. Given that

$$\begin{aligned}
5^{k+1} - 1 &= (4l + 1)5 - 1 \\
&= 4(5l) + 5 - 1 \\
&= 4(5l) + 4 \\
&= 4(5l + 1),
\end{aligned}$$

for $5l + 1$ an integer, it follows that $4|(5^{k+1} - 1)$. This shows that the implication $P(k)$ implies $P(k+1)$ is true. □

Before we proceed, note for $n \in \mathbb{N}$ that $n \geq 1$. This implies $n \cdot n \geq n \cdot 1$ or that $n^2 \geq n$. Given that $n \geq 1$, it also follows that $n^2 \geq 1$. These two facts will be needed when proving the next result.

Proposition 4.2.5. *For all $n \in \mathbb{N}$, $4^n > n^2$.*

Proof. Let $P(n)$ denote the statement $4^n > n^2$.

Base Case: We must show that $P(1)$ is true. When $n = 1$, it follows that $4^n = 4^1 = 4$ and $n^2 = 1^2 = 1$. Since $4 > 1$, we have $4^n > n^2$, when $n = 1$, and that the statement $P(1)$ is true.

Inductive Case: Let $k \in \mathbb{N}$ and assume that $P(k)$ is true. Therefore, we have

$$4^k > k^2.$$

We must show $P(k+1)$ is true or that

$$4^{k+1} > (k + 1)^2.$$

First note that $4^{k+1} = 4^1 4^k = 4 \cdot 4^k$. Using the fact that $4^k > k^2$, we get that $4^{k+1} = 4 \cdot 4^k > 4k^2$. Given that $4k^2 = k^2 + k^2 + k^2 + k^2$ with the facts that $k^2 \geq k$ and $k^2 \geq 1$, we get that

$$4k^2 = k^2 + k^2 + k^2 + k^2 \geq k^2 + k + k + 1 = k^2 + 2k + 1 = (k+1)^2.$$

Putting all this together results in

$$4^{k+1} > 4k^2 \geq (k+1)^2$$

or that

$$4^{k+1} > (k+1)^2.$$

This shows that the implication $P(k)$ implies $P(k+1)$ is true.   □

You can also prove statements concerning sets with induction.

Proposition 4.2.6. *Let $A_1, A_2, \ldots, A_n$ be sets, and let $\mathcal{U}$ be the universe. Then for all $n \in \mathbb{N}$,*

$$\overline{A_1 \cap A_2 \cap \cdots \cap A_n} = \overline{A_1} \cup \overline{A_2} \cup \cdots \cup \overline{A_n}.$$

Proof. Let $P(n)$ denote the statement $\overline{A_1 \cap A_2 \cap \cdots \cap A_n} = \overline{A_1} \cup \overline{A_2} \cup \cdots \cup \overline{A_n}$.

Base Case: We must show that $P(1)$ is true. When $n = 1$, the statement $\overline{A_1} = \overline{A_1}$ is true by Lemma 3.2.4 (i).

Inductive Case: Let $k \in \mathbb{N}$ and assume that $P(k)$ is true. This means

$$\overline{A_1 \cap A_2 \cap \cdots \cap A_k} = \overline{A_1} \cup \overline{A_2} \cup \cdots \cup \overline{A_k}.$$

We must show $P(k+1)$ is true or that

$$\overline{A_1 \cap A_2 \cap \cdots \cap A_k \cap A_{k+1}} = \overline{A_1} \cup \overline{A_2} \cup \cdots \cup \overline{A_{k+1}}.$$

First note that

$$\overline{A_1 \cap A_2 \cap \cdots \cap A_k \cap A_{k+1}} = \overline{(A_1 \cap A_2 \cap \cdots \cap A_k) \cap A_{k+1}}.$$

Applying Theorem 3.3.7 (iii) to the two sets $A_1 \cap A_2 \cap \cdots \cap A_k$ and A_{k+1} results in

$$\overline{A_1 \cap A_2 \cap \cdots \cap A_{k+1}} = \overline{(A_1 \cap A_2 \cap \cdots \cap A_k) \cap A_{k+1}}$$
$$= \overline{A_1 \cap A_2 \cap \cdots \cap A_k} \cup \overline{A_{k+1}}.$$

By the induction hypothesis, we know that

$$\overline{A_1 \cap A_2 \cap \cdots \cap A_k} = \overline{A_1} \cup \overline{A_2} \cup \cdots \cup \overline{A_k}.$$

Using it, we obtain

$$\begin{aligned}
\overline{A_1 \cap A_2 \cap \cdots \cap A_{k+1}} &= \overline{A_1 \cap A_2 \cap \cdots \cap A_k \cup A_{k+1}} \\
&= (\overline{A_1} \cup \overline{A_2} \cup \cdots \cup \overline{A_k}) \cup \overline{A_{k+1}} \\
&= \overline{A_1} \cup \overline{A_2} \cup \cdots \cup \overline{A_{k+1}}.
\end{aligned}$$

This shows that the implication $P(k)$ implies $P(k+1)$ is true. □

We now need to examine a more general principle of mathematical induction. There are a number of statements in mathematics that instead of saying, "$(\forall n \in \mathbb{N}), P(n)$," say, "$P(n)$ is true for all $n \in \mathbb{Z}$, such that $n \geq n_0$ for some integer n_0." The following statements are two examples.

- For all integers $n \geq 0$, $1 + 2 + 2^2 + \cdots + 2^n = 2^{n+1} - 1$.

- For all integers $n \geq 5$, $2^n > n^2$.

To prove these statements, we need a more general form of the Principal of Mathematical Induction.

Theorem 4.2.7 (Principle of Mathematical Induction-Version 2). *Let $P(n)$ be a statement for every integer n with $n \geq n_0$. If both*

(i) *$P(n_0)$ is true and*

(ii) *the implication $P(k)$ implies $P(k+1)$ is true for all $k \geq n_0$,*

then $P(n)$ is true for all $n \geq n_0$.

All of the terminology introduced earlier remains the same.

Proposition 4.2.8. *For all integers $n \geq 0$, $1+2+2^2+\cdots+2^n = 2^{n+1}-1$.*

Proof. Let $P(n)$ denote the statement $1 + 2 + 2^2 + \cdots + 2^n = 2^{n+1} - 1$.

Base Case: We must show that $P(0)$ is true. When $n = 0$, the sum is 1 and $2^{n+1} - 1 = 2^{0+1} - 1 = 2 - 1 = 1$. Since the resulting values are equal, $P(0)$ is true.

Inductive Case: Let k be an integer, such that $k \geq 0$, and assume that $P(k)$ is true. This means

$$1 + 2 + 2^2 + \cdots + 2^k = 2^{k+1} - 1.$$

We must show $P(k+1)$ is true or that

$$1 + 2 + 2^2 + \cdots + 2^k + 2^{k+1} = 2^{(k+1)+1} - 1 = 2^{k+2} - 1.$$

The calculation

$$
\begin{aligned}
1 + 2 + 2^2 + \cdots + 2^k + 2^{k+1} &= (1 + 2 + 2^2 + \cdots + 2^k) + 2^{k+1} \\
&= 2^{k+1} - 1 + 2^{k+1} \\
&= 2 \cdot 2^{k+1} - 1 \\
&= 2^{k+2} - 1
\end{aligned}
$$

shows that the implication $P(k)$ implies $P(k+1)$ is true. □

Proposition 4.2.9. *For all integers* $n \geq 5$, $2^n > n^2$.

Proof. Let $P(n)$ denote the statement $2^n > n^2$.

Base Case: We must show that $P(5)$ is true. When $n = 5$, it follows that $2^n = 2^5 = 32$ and $n^2 = 5^2 = 25$. Since $32 > 25$, we have that $P(5)$ is true.

Inductive Case: Let k be an integer, such that $k \geq 5$, and assume that $P(k)$ is true. It follows that

$$2^k > k^2.$$

We must show $P(k+1)$ is true or that

$$2^{k+1} > (k+1)^2.$$

We start by observing that $2^{k+1} = 2^1 2^k = 2 \cdot 2^k$. Since $2^k > k^2$, we have that $2^{k+1} = 2 \cdot 2^k > 2k^2 = k^2 + k^2$. Given that $k \geq 5$, it follows that $k^2 \geq 5k = 4k + k$. The fact that $4 > 2$ implies $4k > 2k$. Furthermore, $k \geq 5 > 1$ means that $k > 1$. Putting all this together, we have that

$$k^2 + k^2 \geq k^2 + 5k = k^2 + 4k + k > k^2 + 2k + 1 = (k+1)^2.$$

This results in $2^{k+1} > 2k^2 > (k+1)^2$ or more simply that $2^{k+1} > (k+1)^2$. This shows that the implication $P(k)$ implies $P(k+1)$ is true. □

An important inequality that will be needed later is stated here. Its proof, using induction, is an exercise.

Lemma 4.2.10. *For all* $n \in \mathbb{Z}$ *with* $n \geq 0$, $2^n > n$.

Exercises 4.2: In all cases, prove by mathematical induction.

(1) Prove that $2 + 4 + 6 + \cdots + 2n = n(n + 1)$ for all $n \in \mathbb{N}$.

(2) Prove that $3 + 6 + 9 + \cdots + 3n = \frac{3n(n+1)}{2}$ for all $n \in \mathbb{N}$.

(3) Prove that $8 + 16 + 24 + \cdots + 8n = 4n(n + 1)$ for all $n \in \mathbb{N}$.

(4) Prove that $4 + 10 + 16 + \cdots + 6n - 2 = n(3n + 1)$ for all $n \in \mathbb{N}$.

(5) Prove that $2 + 7 + 12 + \cdots + 5n - 3 = \frac{n(5n-1)}{2}$ for all $n \in \mathbb{N}$.

(6) Prove that $1^3 + 2^3 + 3^3 + \cdots + n^3 = \frac{n^2(n+1)^2}{4}$ for all $n \in \mathbb{N}$.

(7) Prove that $2^2 + 4^2 + 6^2 + \cdots + (2n)^2 = \frac{2n(2n+1)(2n+2)}{6}$ for all $n \in \mathbb{N}$.

(8) For $n \in \mathbb{N}$, prove that $1{\cdot}2 + 2{\cdot}3 + 3{\cdot}4 + \cdots + n(n+1) = \frac{n(n+1)(n+2)}{3}$.

(9) For $n \in \mathbb{N}$, prove that $1{\cdot}3 + 2{\cdot}4 + 3{\cdot}5 + \cdots + n(n+2) = \frac{n(n+1)(2n+7)}{6}$.

(10) Prove that $\frac{1}{1{\cdot}2} + \frac{1}{2{\cdot}3} + \frac{1}{3{\cdot}4} + \cdots + \frac{1}{n(n+1)} = \frac{n}{n+1}$ for all $n \in \mathbb{N}$.

(11) Prove that $5 | (2^{4n-2} + 1)$ for all $n \in \mathbb{N}$.

(12) Prove that $3 | (4^n - 1)$ for all $n \in \mathbb{N}$.

(13) Prove that $3 | (n^3 - n)$ for all $n \in \mathbb{N}$.

(14) Prove that $5 | (7^n - 2^n)$ for all $n \in \mathbb{N}$.

(15) Prove that $6 | (3n(n + 1))$ for all $n \in \mathbb{N}$.

(16) Prove Lemma 4.2.10: for all $n \in \mathbb{Z}$ with $n \geq 0$, $2^n > n$.

(17) Prove that $3^n > n2^n$ for all $n \in \mathbb{N}$.

(18) Prove for sets $A_1, A_2, \ldots, A_n$, with $n \in \mathbb{N}$, and the universe $\mathcal{U}$ that
$$\overline{A_1 \cup A_2 \cup \cdots \cup A_n} = \overline{A_1} \cap \overline{A_2} \cap \cdots \cap \overline{A_n}.$$

(19) Prove for sets $A_1, A_2, \ldots, A_n$, with $n \in \mathbb{N}$, and set B that
$$B - (A_1 \cup A_2 \cup \cdots \cup A_n) = (B - A_1) \cap (B - A_2) \cap \cdots \cap (B - A_n).$$

(20) Prove for sets $A_1, A_2, \ldots, A_n$, with $n \in \mathbb{N}$, and set B that
$$B - (A_1 \cap A_2 \cap \cdots \cap A_n) = (B - A_1) \cup (B - A_2) \cup \cdots \cup (B - A_n).$$

(21) Prove $1 + 3 + 3^2 + \cdots + 3^n = \frac{3^{n+1}-1}{2}$ for all integers n with $n \geq 0$.

(22) Prove $1 + 5 + 5^2 + \cdots + 5^n = \frac{5^{n+1}-1}{4}$ for all integers n with $n \geq 0$.

(23) Prove $1 + r + r^2 + \cdots + r^n = \frac{1-r^{n+1}}{1-r}$ for all integers n with $n \geq 0$ and $r \in \mathbb{R}$ with $r \neq 1$.

(24) Prove that $3^n > n^2$ for all integers n with $n \geq 2$.

(25) Prove that $4^n > 3^n + 2$ for all integers n with $n \geq 2$.

(26) Prove that $(n+1)^2 \geq n^2 + 1$ for all integers n with $n \geq 0$.

(27) Prove that $n^2 - n - 6 > 0$ for all integers n with $n \geq 4$.

4.3 PROOF BY STRONG INDUCTION

There are a number of statements in mathematics that require a different, ostensibly stronger form of induction to prove them. To motivate it, consider recursively defined sequences. In such a sequence, the first n_0 terms of the sequence, for $n_0 \in \mathbb{N}$, are explicitly given and every n^{th} term of the sequence, for $n \in \mathbb{N}$ with $n > n_0$, is obtained via an expression involving previous terms.

For example, consider the sequence a_n, for $n \in \mathbb{N}$, defined by

$$a_1 = 2$$
$$a_2 = 6$$
$$a_n = 2a_{n-1} - a_{n-2} \text{ for } n \geq 3.$$

After writing out the first few terms

$$a_1 = 2, a_2 = 6, a_3 = 10, a_4 = 14, a_5 = 18, a_6 = 22,$$

the pattern indicates that $a_n = 4n - 2$. The fact that $a_n = 4n - 2$ is referred to as the **closed-form** formula for the recursively defined sequence. Now suppose we tried to prove this fact using the form of induction we currently know. Showing the statement is true for $n = 1$ is simple, but we would run into trouble at the inductive case. We assume the statement is true for $k \geq 1$ or that $a_k = 4k - 2$. We would then try to prove the statement is true for $k+1$ or that $a_{k+1} = 4(k+1) - 2$. We know from the definition of the sequence that $a_{k+1} = 2a_{(k+1)-1} - a_{(k+1)-2} = 2a_k - a_{k-1}$. But at this point we are stuck. We know $a_k = 4k - 2$ but we have no information on a_{k-1}. We need the following form of induction.

Theorem 4.3.1 (The Principle of Strong Mathematical Induction). *Let $P(n)$ be a statement for every $n \in \mathbb{N}$. If both*

(i) $P(1), P(2), \ldots, P(m)$ *are true for $m \geq 1$ and*

(ii) *the implication $P(i)$, for all integers i such that $1 \leq i \leq k$, implies $P(k+1)$ is true for all $k \in \mathbb{N}$ with $k \geq m$,*

then $P(n)$ is true for all $n \in \mathbb{N}$.

First note that the base case may require showing that the statement $P(n)$ is true for more than one initial value. Second, in the inductive case, instead of assuming the statement is true for k, you assume the statement is true for all natural numbers less than or equal to k. This gives you the information needed to prove that $P(k+1)$ is true. It should also be noted that the Principle of Mathematical Induction and the Principle of Strong Mathematical Induction are equivalent.

Proposition 4.3.2. *For the sequence defined by*

$$a_1 = 2$$
$$a_2 = 6$$
$$a_n = 2a_{n-1} - a_{n-2} \text{ for } n \geq 3,$$

the closed-form formula is $a_n = 4n - 2$ for all $n \in \mathbb{N}$.

Proof. Let $P(n)$ denote the statement $a_n = 4n - 2$.

Base Case: We must show that $P(1)$ and $P(2)$ are true. When $n = 1$, we have $a_1 = 2$ and $4n - 2 = 4 \cdot 1 - 2 = 2$. When $n = 2$, we have $a_2 = 6$ and $4n - 2 = 4 \cdot 2 - 2 = 6$. Thus $P(n)$ is true for $n = 1$ and $n = 2$.

Inductive Case: Let k be an integer, such that $k \geq 2$, and assume that $P(i)$ is true for all integers i such that $1 \leq i \leq k$. This means

$$a_i = 4i - 2.$$

for $1 \leq i \leq k$. We must show $P(k+1)$ is true or that

$$a_{k+1} = 4(k+1) - 2.$$

By the definition of the sequence, we start by observing that

$$a_{k+1} = 2a_{(k+1)-1} - a_{(k+1)-2} = 2a_k - a_{k-1}.$$

Since $1 \leq k-1, k \leq k$, we have that $a_k = 4k-2$ and $a_{k-1} = 4(k-1)-2$. This implies

$$
\begin{aligned}
a_{k+1} &= 2a_k - a_{k-1} \\
&= 2(4k-2) - (4(k-1)-2) \\
&= 8k - 4 - 4k + 4 + 2 \\
&= 4k + 2 \\
&= 4k + 4 - 2 \\
&= 4(k+1) - 2,
\end{aligned}
$$

showing that the implication $P(i)$, for $1 \leq i \leq k$, implies $P(k+1)$ is true. $\qquad\square$

Proposition 4.3.3. *For the sequence defined by*

$$
\begin{aligned}
a_1 &= 1 \\
a_2 &= 3 \\
a_3 &= 7 \\
a_n &= a_{n-1} + a_{n-2} - a_{n-3} + 4 \ \textit{for } n \geq 4,
\end{aligned}
$$

the closed-form formula is $a_n = n^2 - n + 1$ for all $n \in \mathbb{N}$.

Proof. Let $P(n)$ denote the statement $a_n = n^2 - n + 1$.

Base Case: We must show that $P(1), P(2)$, and $P(3)$ are true. When $n = 1$, we have $a_1 = 1$ and $n^2 - n + 1 = 1^2 - 1 + 1 = 1$. When $n = 2$, we have $a_2 = 3$ and $n^2 - n + 1 = 2^2 - 2 + 1 = 3$. And finally, when $n = 3$, we have $a_3 = 7$ and $n^2 - n + 1 = 3^2 - 3 + 1 = 7$. Thus $P(n)$ is true for $n = 1, n = 2$, and $n = 3$.

Inductive Case: Let k be an integer, such that $k \geq 3$, and assume that $P(i)$ is true for all integers i with $1 \leq i \leq k$. Thus we have

$$
a_i = i^2 - i + 1
$$

for $1 \leq i \leq k$. We must show $P(k+1)$ is true or that

$$
a_{k+1} = (k+1)^2 - (k+1) + 1.
$$

By the definition of the sequence, we start by observing that

$$
a_{k+1} = a_{(k+1)-1} + a_{(k+1)-2} - a_{(k+1)-3} + 4 = a_k + a_{k-1} - a_{k-2} + 4.
$$

Since $1 \leq k-2, k-1, k \leq k$, we have that

$$a_k = k^2 - k + 1,$$
$$a_{k-1} = (k-1)^2 - (k-1) + 1 = k^2 - 3k + 3, \text{ and}$$
$$a_{k-2} = (k-2)^2 - (k-2) + 1 = k^2 - 5k + 7.$$

This implies

$$
\begin{aligned}
a_{k+1} &= a_k + a_{k-1} - a_{k-2} + 4 \\
&= k^2 - k + 1 + k^2 - 3k + 3 - [k^2 - 5k + 7] + 4 \\
&= k^2 + k + 1 \\
&= k^2 + 2k + 1 - k - 1 + 1 \\
&= (k+1)^2 - (k+1) + 1,
\end{aligned}
$$

showing that the implication $P(i)$, for $1 \leq i \leq k$, implies $P(k+1)$ is true. □

One of the most famous recursively defined sequences in mathematics is the Fibonacci sequence. It is defined as follows:

$$F_1 = 1$$
$$F_2 = 1$$
$$F_n = F_{n-1} + F_{n-2} \text{ for } n \geq 3$$

The Fibonnaci sequence is

$$1, 1, 2, 3, 5, 8, 13, 21, 34, 55, 89, 144, 233, \ldots$$

Credit for this sequence goes to Italian mathematician Leonardo Pisano (ca. 1170–1250), whose mathematical pen name was Fibonacci. It appears in a number of places in mathematics. If you are interested in exploring the Fibonacci sequence further, I suggest the book [PL07]. The fact that there is a closed-form formula for the Fibonacci sequence is not obvious. But there is one and it involves the Golden Ratio $\phi = \frac{1+\sqrt{5}}{2}$.

Theorem 4.3.4. *For the Fibonacci sequence defined by*

$$F_1 = 1$$
$$F_2 = 1$$
$$F_n = F_{n-1} + F_{n-2} \text{ for } n \geq 3,$$

the closed-form formula is $F_n = \frac{r^n - s^n}{r - s}$, *for all* $n \in \mathbb{N}$, *with* $r = \frac{1+\sqrt{5}}{2}$ *and* $s = \frac{1-\sqrt{5}}{2}$.

The proof of Theorem 4.3.4 is left as an exercise.

Two important results concerning prime numbers, mentioned in Chapter 2, will be stated here, as the proof of the first result requires strong induction and the proof of the second result requires the first. Recall that a positive integer greater than 1 is prime if its only positive divisors are 1 and itself.

Theorem 4.3.5. *Any integer greater than or equal to 2 has at least one prime divisor.*

Proof. Let $n \in \mathbb{N}$ with $n \geq 2$, and let $P(n)$ be the statement that n has at least one prime divisor.

Base Case: Given that 2 is prime, it is clear that 2 has a prime divisor.

Inductive Case: Let k be an integer, such that $k \geq 2$, and assume that $P(i)$ is true for all integers i with $2 \leq i \leq k$. It follows that each integer i, where $2 \leq i \leq k$, has at least one prime divisor.

We must show $P(k+1)$ is true or that $k+1$ has at least one prime divisor. If $k+1$ is prime, then $k+1$ is the prime divisor of $k+1$ and we are done. Suppose that $k+1$ is not prime. Then $k+1$ is composite and $k+1 = ab$ for integers a and b with $2 \leq a, b \leq k$. Given that $2 \leq a \leq k$, we know that a has at least one prime divisor or that $a = pl$ for some integer l, with $l \geq 1$, and prime p. This means that

$$k + 1 = ab = (pl)b = p(lb).$$

Since lb is an integer, p divides $k+1$ or $k+1$ has a prime divisor. This shows that the implication $P(i)$, for $2 \leq i \leq k$, implies $P(k+1)$ is true. □

The proof of the next result does not require strong induction, but does require Theorem 4.3.5.

Theorem 4.3.6. *There are infinitely many prime numbers.*

Proof. Suppose that there are only finitely many prime numbers $p_1, p_2, \ldots, p_n$ for $n \in \mathbb{N}$. Now consider the integer $m = p_1 p_2 \cdots p_n + 1$. Suppose that m is prime. Given that $m > p_i$, for $i = 1, 2, \ldots, n$, it is a prime number not in the list $p_1, p_2, \ldots, p_n$. This is a contradiction.

Now suppose that m is not prime. By Theorem 4.3.5, we know that m is divisible by at least one prime. Assume without loss of generality that p_1 divides m. This means that $m = lp_1$ for some integer l or that

$$lp_1 = p_1 p_2 \cdots p_n + 1.$$

This implies

$$lp_1 - p_1p_2\cdots p_n = 1$$

or that

$$p_1(l - p_2\cdot p_n) = 1,$$

where $l - p_2\cdots p_n$ is an integer. This means that p_1 divides 1, which is also a contradiction. Thus there are infinitely many primes. □

Exercises 4.3

(1) For the sequence defined by

$$a_1 = 1$$
$$a_n = a_{n-1} + 5 \text{ for } n \geq 3,$$

prove that the closed-form formula is $a_n = 5n - 4$ for all $n \in \mathbb{N}$.

(2) For the sequence defined by

$$a_1 = 3$$
$$a_2 = 6$$
$$a_n = 2a_{n-1} - a_{n-2} \text{ for } n \geq 3,$$

prove that the closed-form formula is $a_n = 3n$ for all $n \in \mathbb{N}$.

(3) For the sequence defined by

$$a_1 = 2$$
$$a_2 = 4$$
$$a_n = 5a_{n-1} - 6a_{n-2} \text{ for } n \geq 3,$$

prove that the closed-form formula is $a_n = 2^n$ for all $n \in \mathbb{N}$.

(4) For the sequence defined by

$$a_1 = 1$$
$$a_2 = 4$$
$$a_n = 2a_{n-1} - a_{n-2} + 2 \text{ for } n \geq 3,$$

prove that the closed-form formula is $a_n = n^2$ for all $n \in \mathbb{N}$.

(5) For the sequence defined by

$$a_1 = -1$$
$$a_2 = 2$$
$$a_n = -2a_{n-1} - a_{n-2} \text{ for } n \geq 3,$$

prove that the closed-form formula is $a_n = (-1)^n n$ for all $n \in \mathbb{N}$.

(6) For the sequence defined by

$$a_1 = 1$$
$$a_2 = 4$$
$$a_3 = 9$$
$$a_n = a_{n-1} - a_{n-2} + a_{n-3} + 4n - 6 \text{ for } n \geq 4,$$

prove that the closed-form formula is $a_n = n^2$ for all $n \in \mathbb{N}$.

(7) Prove Theorem 4.3.4: for the Fibonacci sequence defined by

$$F_1 = 1$$
$$F_2 = 1$$
$$F_n = F_{n-1} + F_{n-2} \text{ for } n \geq 3,$$

the term $F_n = \frac{r^n - s^n}{r - s}$ for all $n \in \mathbb{N}$ with $r = \frac{1+\sqrt{5}}{2}$ and $s = \frac{1-\sqrt{5}}{2}$.

(8) Prove that there are infinitely many positive even integers.

(9) Prove the existence part of Theorem 2.3.7: every positive integer greater than 1 is prime or can be written as a product of finitely many primes. (The proof of uniqueness requires a few concepts not introduced here and can be found in [JJ98].)

CHAPTER **5**

Relations

5.1 INTRODUCTION

Given a set of numbers, there are a variety of ways to compare or relate the numbers in that set. With the real numbers $\mathbb{R}$, the most commonly used relations are $<, \leq, =, \neq, >$, and $\geq$. Since $-2 < 6$, we have that -2 is related to 6 via less than. And given that $5 > -1$, we know 5 is related to -1 via greater than. We even define collections of numbers using these comparisons. For example, a real number r is negative if $r < 0$ and positive if $r > 0$.

The concept of divides also provides a method for relating integers. In other words a is related to b if and only if $a|b$, for all $a, b \in \mathbb{Z}$. The integers 3 and 12 are related this way as $3|12$, but 3 would not be related to 8 as $3 \nmid 8$.

These types of comparisons are not limited to sets of numbers. For example, take the set of the words in the dictionary. One way to relate two words is that the first word is related to the second word if the first word comes alphabetically before the second word (lexicographic order).

Now on to a formal definition.

Definition 5.1.1. *Let A and B be sets. A **relation** R from A to B is a subset of $A \times B$. If $A = B$, then R is a **relation on** A.*

If the ordered pair (a, b) is in the relation R from A to B ($(a, b) \in R$), we say that "a is related to b" and often denote this by $a\,R\,b$. If (a, b) is not in the relation R from A to B ($(a, b) \notin R$), we say that "a is not related to b" and often denote this by $a\,\not\!R\,b$.

Example 5.1.2.

(i) *Consider the set $A = \{1, 2, 3, 4, 5, 6\}$ and $B = \{2, 4, 9\}$. Then*

$$R = \{(1, 2), (1, 4), (1, 9), (2, 4), (3, 2), (4, 9), (5, 2)\}$$

is an example of a relation from A to B. Here $(1, 4) \in R$ or $1\,R\,4$, but $(6, 2) \notin R$ or $6\,R\!\!\!/\,2$. The element 1 in A is related to every element in B, but 6 in A is not related to any element in B.

(ii) *Consider the set $A = \{a, b, c, d, e\}$. Then*

$$R = \{(a, a), (a, d), (b, c), (b, e), (c, a), (d, a), (d, b)\}$$

and

$$S = \{(a, b), (b, c), (c, d), (d, e), (e, a)\}$$

are both relations on A. Each relation is simply defined by the collection of ordered pairs.

(iii) *Consider the set $A = \{1, 2, 3, 4\}$. Define a relation R on A by $a\,R\,b$ if and only if $a|b$ for $a, b \in A$. Here, we have $2\,R\,4$ as $2|4$, but $2\,R\!\!\!/\,3$ as $2 \nmid 3$. By examining all possible pairs of elements, we get that*

$$R = \{(1, 1), (1, 2), (1, 3), (1, 4), (2, 2), (2, 4), (3, 3), (4, 4)\}.$$

(iv) *The set $R = \{(x, y) \in \mathbb{Z} \times \mathbb{Z} : x + y \geq 5\}$ defines a relation on $\mathbb{Z}$ containing infinitely many ordered pairs. For example, we have $(3, 2) \in R$ or $3\,R\,2$ as $3 + 2 \geq 5$ and $(-3, 9) \in R$ or $-3\,R\,9$ as $-3 + 9 \geq 5$. However, $(1, 3) \notin R$ or $1\,R\!\!\!/\,3$ since $1 + 3 \ngeq 5$. Geometrically, every point in the Cartesian plane with integer coordinates on or above the line $x + y = 5$ is in R.*

(v) *The set $S = \{(x, y) \in \mathbb{R} \times \mathbb{R} : x^2 + y^2 < 1\}$ defines a relation on $\mathbb{R}$. For this relation, it follows that $\frac{1}{3}\,R\,\frac{-1}{2}$ as $\left(\frac{1}{3}\right)^2 + \left(\frac{-1}{2}\right)^2 = \frac{13}{36} < 1$ and $\sqrt{2}\,R\!\!\!/\,-1$ as $\left(\sqrt{2}\right)^2 + (-1)^2 = 3 \geq 1$. Geometrically, the relation S is the set of points in the Cartesian plane inside the unit circle.*

Definition 5.1.3. *Let R be a relation from a set A to a set B. The **inverse relation** of R, denoted by R^{-1} is the relation from B to A defined by*

$$R^{-1} = \{(b, a) : (a, b) \in R\}.$$

Let $A = \{1, 2, 3, 4, 5, 6\}$ and $B = \{2, 4, 9\}$. Recall the relation

$$R = \{(1, 2), (1, 4), (1, 9), (2, 4), (3, 2), (4, 9), (5, 2)\}$$

from A to B defined in Example 5.1.2 (i). Then R^{-1} is a relation from B to A where

$$R^{-1} = \{(2, 1), (4, 1), (9, 1), (4, 2), (2, 3), (9, 4), (2, 5)\}.$$

Exercises 5.1

(1) Let $A = \{1, 2, 3, 4, 5\}$ and $B = \{6, 8, 10\}$. Write out the relation R from A to B defined by $a\,R\,b$ if and only if $2a < b$ for $a \in A$ and $b \in B$. Determine the relation R^{-1} from B to A.

(2) Let $A = \left\{0, \frac{1}{2}, \sqrt{2}, \pi, e\right\}$. Write out the relation R on A defined by $a > b$ for $a, b \in A$. Determine the relation R^{-1} on A.

(3) Let $A = \{1, 2, 3, 4, 6, 12\}$. Write out the relation R on A defined by $a\,R\,b$ if and only if $a|b$ for $a, b \in A$. Find the relation R^{-1} on A.

(4) Let $A = \{-4, -2, -1, 0, 1, 2, 4\}$. Write out the relation R on A defined by $a\,R\,b$ if and only if $b = a^2$ for $a, b \in A$.

(5) Let $A = \{0, 1, 2, 3, 4, 5, 6, 7, 8\}$. Write out the relation R on A defined by $a\,R\,b$ if and only if $a \equiv b \pmod{3}$ for $a, b \in A$.

(6) Consider the relation R on $\mathbb{Z}$ defined by $a\,R\,b$ if and only if $a \equiv b \pmod{3}$.

    (a) Determine the set of all elements $x \in \mathbb{Z}$ such that $x\,R\,0$.

    (b) Determine the set of all elements $y \in \mathbb{Z}$ such that $y\,R\,1$.

    (c) Determine the set of all elements $z \in \mathbb{Z}$ such that $z\,R\,2$.

(7) Consider the relation R on $\mathbb{Z}$ defined by $a\,R\,b$ if and only if $a + b$ is even for $a, b \in \mathbb{Z}$.

    (a) List eight ordered pairs in R.

    (b) Determine the set of all elements $x \in \mathbb{Z}$ such that $x\,R\,0$.

    (c) Determine the set of all elements $y \in \mathbb{Z}$ such that $y\,R\,1$.

(8) Consider the relation R on $\mathbb{R}$ defined by $a\,R\,b$ if and only if $ab \in \mathbb{Q}$ for $a, b \in \mathbb{R}$.

  (a) List ten ordered pairs in R.

  (b) Determine the set of all elements $x \in \mathbb{R}$ such that $x \, R \, 0$.

(9) Consider the relation R on $\mathbb{N}$ defined by $a \, R \, b$ if and only if $b = a + 2$ and both a and b are prime for $a, b \in \mathbb{R}$. List five elements in R.

(10) Consider the relation R on $\mathbb{N}$ defined by $a \, R \, b$ if and only if there exists an integer $c \in \mathbb{N}$ such that $a^2 + b^2 = c^2$ for $a, b \in \mathbb{N}$. List five elements in R.

(11) Consider the relation R on $\mathbb{R}$ defined by $x \, R \, y$ if and only if $y = x + \frac{1}{x}$ for $x, y \in \mathbb{R}$. List 10 elements in R.

(12) Let A be a nonempty set. Describe two distinct relations R on A.

(13) Let $A = \{1, 2\}$. Write all of the relations that can exist on A.

5.2 PROPERTIES OF RELATIONS

There are a number of properties that a relation on a set A can satisfy. We are interested in the following three.

Definition 5.2.1. *Let R be a relation on a set A.*

  (i) *The relation R is **reflexive** if and only if $a \, R \, a$ for every element $a \in A$.*

  (ii) *The relation R is **symmetric** if and only if $a \, R \, b$ implies $b \, R \, a$ for all $a, b \in A$.*

  (iii) *The relation R is **transitive** if and only if $a \, R \, b$ and $b \, R \, c$ implies $a \, R \, c$ for all $a, b, c \in A$.*

Example 5.2.2. *Consider the set $A = \{0, 1, 2, 3, 4\}$ and the following relations on A.*

  (i) $R_1 = \{(0,0), (0,1), (1,0), (2,2), (3,4), (4,2)\}$

  While $0 \, R_1 \, 0$ and $2 \, R_1 \, 2$, this relation is not reflexive since $1 \, \not R_1 \, 1$, $3 \, \not R_1 \, 3$, and $4 \, \not R_1 \, 4$. This relation is not symmetric as $3 \, R_1 \, 4$, but $4 \, \not R_1 \, 3$. This relation is not transitive as $3 \, R_1 \, 4$ and $4 \, R_1 \, 2$, but $3 \, \not R_1 \, 2$.

(ii) $R_2 = \{(0,0), (1,1), (2,2), (2,3), (2,4), (3,3), (3,4), (4,4)\}$

This relation is reflexive as $a\,R_2\,a$ for all $a \in A$. This relation is not symmetric as $2\,R_2\,3$, but $3\,\cancel{R_2}\,2$. This relation is transitive as in every case where $a\,R_2\,b$ and $b\,R_2\,c$, we also have $a\,R_2\,c$ for $a, b, c \in A$. For example, $2\,R_2\,3$ and $3\,R_2\,4$, and we see $2\,R_2\,4$.

(iii) $R_3 = \{(0,0), (1,1), (2,2), (2,3), (2,4), (3,2), (3,3), (3,4), (4,2),$
$\qquad (4,3), (4,4)\}$

This relation is reflexive as $a\,R_3\,a$ for all $a \in A$. This relation is symmetric as in every case where $a\,R_3\,b$, you also have $b\,R_3\,a$. For example, $2\,R_3\,4$ and we find $4\,R_3\,2$. This relation is transitive as in every case where $a\,R_3\,b$ and $b\,R_3\,c$, we find $a\,R_3\,c$ for all $a, b, c \in A$.

(iv) $R_4 = \{(0,1)\}$

This relation is not reflexive as not every element in A is related to itself. For example, 0 is not related to 0. This relation is not symmetric as $0\,R_4\,1$, but $1\,\cancel{R_4}\,0$. This relation is transitive. This follows from that fact that there do not exist elements $a, b, c \in A$ such that $a\,R_4\,b$ and $b\,R_4\,c$. Since the hypothesis of the statement if $a\,R_4\,b$ and $b\,R_4\,c$ implies $a\,R_4\,c$ is false, the relation is vacuously transitive.

(v) $R_5 = \{(0,0)\}$

This relation is not reflexive as not every element in A is related to itself. For example, 1 is not related to 1. This relation is symmetric as $0\,R_5\,0$ implies $0\,R_5\,0$. This relation is transitive since you have $0\,R_5\,0$ and $0\,R_5\,0$, and we see $0\,R_5\,0$.

Now on to relations on an infinite set.

Example 5.2.3. *Consider the following relations on $\mathbb{Z}$.*

(i) Define relation S on $\mathbb{Z}$ by $a\,S\,b$ if and only if $a \geq b$ for $a, b \in \mathbb{Z}$.

This relation is reflexive as every element $a \in A$ does satisfy $a \geq a$. This relation is not symmetric as $a \geq b$ does not imply $b \geq a$ for every $a, b \in \mathbb{Z}$. For example, $4 \geq 2$ but $2 \not\geq 4$. This relation is transitive as $a \geq b$ and $b \geq c$ implies $a \geq c$ for all $a, b, c \in \mathbb{Z}$.

(ii) Define relation T on $\mathbb{Z}$ by $a\,T\,b$ if and only if $a = b$ for $a, b \in \mathbb{Z}$.

This relation is reflexive as every element $a \in A$ does satisfy $a = a$.

This relation is symmetric as $a = b$ does imply $b = a$ for every $a, b \in \mathbb{Z}$. This relation is transitive as $a = b$ and $b = c$ implies $a = c$ for all $a, b, c \in \mathbb{Z}$. Written out, this relation is

$$T = \{\dots, (-2, -2), (-1, -1), (0, 0), (1, 1), (2, 2), \dots\}.$$

(iii) *Define relation U on $\mathbb{Z}$ by $a \, U \, b$ if and only if $a|b$ for $a, b \in \mathbb{Z}$.*

Let a be an element of $\mathbb{Z}$. Given that $a = 1{\cdot}a$, with 1 an integer, we have that $a|a$ and $a \, U \, a$. Thus this relation is reflexive as every element $a \in \mathbb{Z}$ does satisfy $a|a$. This relation is not symmetric as $a|b$ does not imply $b|a$ for every $a, b \in \mathbb{Z}$. For example, $3 \, U \, 12$ as $3|12$, but $12 \, \cancel{U} \, 3$ as $12 \nmid 3$. Now let $a, b, c \in \mathbb{Z}$ such that $a|b$ and $b|c$. It follows from Lemma 2.4.3 that $a|c$. Therefore this relation is transitive as $a|b$ and $b|c$ implies $a|c$ for all $a, b, c \in \mathbb{Z}$.

Exercises 5.2

(1) Determine if each of the following relations on set $A = \{-2, -1, 0, 1, 2\}$ are reflexive, symmetric, or transitive. Justify your answer.

  (a) $R_1 = \{(-2, -2), (0, 0), (0, 2), (2, 0), (2, 1), (2, 2)\}$

  (b) $R_2 = \{(-2, -2), (-1, -1), (0, 0), (0, 1), (1, 1), (1, 2), (2, 2)\}$

  (c) $R_3 = \{(1, 2), (2, 2)\}$

  (d) $R_4 = \{(-1, -1), (0, 1), (1, 0)\}$

  (e) $R_5 = \{(-2, -2), (-1, -1), (-1, 0), (0, -1), (0, 0), (0, 2), (1, 1),$
        $(2, 0), (2, 2)\}$

  (f) $R_6 = \{(-2, -2), (-1, -1), (0, 0), (0, 1), (0, 2), (1, 1), (2, 2)\}$

  (g) $R_7 = \{(0, 0), (0, 1), (1, 0), (1, 1)\}$

  (h) $R_8 = \{(-2, -2), (-1, -1), (0, 0), (0, 1), (0, 2), (1, 0), (1, 1),$
        $(1, 2), (2, 0), (2, 1), (2, 2)\}$

  (i) $R_9 = A \times A$

  (j) $R_{10} = \emptyset$

(2) Determine if the relation defined on the indicated set is reflexive, symmetric, or transitive. Justify your answer.

  (a) The relation R on the set $\mathbb{Z}$ is defined by $a \, R \, b$ if and only if $a \le b$ for $a, b \in \mathbb{Z}$.

(b) The relation R on the set $\mathbb{Z}$ is defined by $a\,R\,b$ if and only if $a + b$ is odd for $a, b \in \mathbb{Z}$.

(c) The relation R on the set $\mathbb{Z}$ is defined by $a\,R\,b$ if and only if $3|(a + b)$ for $a, b \in \mathbb{Z}$.

(d) The relation R on the set $\mathbb{Z}$ is defined by $a\,R\,b$ if and only if $5|(a + 4b)$ for $a, b \in \mathbb{Z}$.

(e) The relation R on the set $\mathbb{Z}$ is defined by $a\,R\,b$ if and only if $a \equiv b \pmod 3$ for $a, b \in \mathbb{Z}$.

(f) The relation R on the set $\mathbb{N}$ is defined by $a\,R\,b$ if and only if $a + b$ is a prime number for $a, b \in \mathbb{Z}$.

(g) The relation R on the set $\mathbb{R}$ is defined by $x\,R\,y$ if and only if $x \leq y + 1$ for $x, y \in \mathbb{R}$.

(h) The relation R on the set $\mathbb{R}$ is defined by $x\,R\,y$ if and only if $xy \geq 0$ for $x, y \in \mathbb{R}$.

(i) Let A be a set, and let $\mathcal{P}(A)$ be the power set of A. The relation R on the set $\mathcal{P}(A)$ is defined by $\alpha\,R\,\beta$ if and only if $\alpha \subseteq \beta$ for $\alpha, \beta \in \mathcal{P}(A)$.

5.3 EQUIVALENCE RELATIONS

We begin this section with a definition.

Definition 5.3.1. *A relation R on a set A is an* **equivalence relation** *if and only if it is reflexive, symmetric, and transitive.*

For the set $A = \{1, 2, 3, 4, 5, 6\}$, the relation

$$R = \{(1,1), (2,2), (2,3), (3,2), (3,3), (4,4), (4,5),$$
$$(4,6), (5,4), (5,5), (5,6), (6,4), (6,5), (6,6)\}$$

on set A is an equivalence relation.

In Exercise 2e from Section 5.2, we saw that the relation R on the set $\mathbb{Z}$ defined by $a\,R\,b$ if and only if $a \equiv b \pmod 3$ for $a, b \in \mathbb{Z}$ is an equivalence relation. More generally, we have the following result.

Proposition 5.3.2. *Let $n \in \mathbb{Z}$ with $n \geq 1$. The relation R on the set $\mathbb{Z}$ defined by $a\,R\,b$ if and only if $a \equiv b \pmod n$ for $a, b \in \mathbb{Z}$ is an equivalence relation.*

Proof. Let $a \in \mathbb{Z}$. To show R is reflexive, we must show that $a\,R\,a$. Given that $a - a = 0 = 0 \cdot n$, with 0 an integer, it follows that $n|(a - a)$. Thus $a \equiv a \pmod{n}$ and $a\,R\,a$.

Now let $a, b \in \mathbb{Z}$ and assume that $a\,R\,b$. This means $a \equiv b \pmod{n}$ or that $n|(a - b)$. Thus there exists an integer l such that $a - b = ln$. To show that R is symmetric, we must show that $b\,R\,a$. Given that

$$
\begin{aligned}
b - a &= (-1)(a - b) \\
&= (-1)ln \\
&= (-l)n,
\end{aligned}
$$

with $-l$ is an integer, it follows that $n|(b - a)$ and that $b \equiv a \pmod{n}$. This results in $b\,R\,a$ and that R is symmetric.

Finally, let $a, b, c \in \mathbb{Z}$ and assume that $a\,R\,b$ and $b\,R\,c$. Therefore, we have $a \equiv b \pmod{n}$ and $b \equiv c \pmod{n}$, meaning $n|(a-b)$ and $n|(b-c)$. Thus there exists integers k and l such that $a - b = kn$ and $b - c = ln$. To show that R is transitive, we must show that $a\,R\,c$. Since

$$
\begin{aligned}
a - c &= a + 0 - c \\
&= a - b + b - c \\
&= (a - b) + (b - c) \\
&= kn - ln \\
&= (k - l)n,
\end{aligned}
$$

with $k - l$ is an integer, this means $n|(a - c)$ and that $a \equiv c \pmod{n}$. Thus relation R is transitive. □

Proposition 5.3.3. *The relation R on set $\mathbb{R}^*$ defined by $x\,R\,y$ if and only if $\frac{x}{y} > 0$, for all $x, y \in \mathbb{R}^*$, is an equivalence relation.*

Proof. Let $x \in \mathbb{R}^*$. To show R is reflexive, we must show that $x\,R\,x$. Given that $\frac{x}{x} = 1 > 0$, this immediately follows.

Now let $x, y \in \mathbb{R}^*$ and assume that $x\,R\,y$. This implies $\frac{x}{y} > 0$. It directly follows that $\frac{y}{x} > 0$. This results in $y\,R\,x$ and that R is symmetric.

Finally, let $x, y, z \in \mathbb{R}^*$ and assume that $x\,R\,y$ and $y\,R\,z$. Thus we have $\frac{x}{y} > 0$ and $\frac{y}{z} > 0$. This implies

$$
\frac{x}{y} \cdot \frac{y}{z} > 0
$$

or that

$$
\frac{x}{z} > 0.
$$

This shows $x \, R \, z$ and that R is transitive. □

Before moving on to properties that equivalence relations have, we need the following definition concerning sets.

Definition 5.3.4. *Let A be a nonempty set. A **partition** $\mathcal{P}$ of A is a set of subsets of A such that*

(i) *for each $X \in \mathcal{P}$, $X \neq \emptyset$,*

(ii) *if X and Y are two distinct elements in $\mathcal{P}$, then $X \cap Y = \emptyset$, and*

(iii) $\bigcup_{X \in \mathcal{P}} X = A$.

Simply put, a partition $\mathcal{P}$ of a nonempty set A is a set of nonempty subsets of A such that any two distinct subsets of A in $\mathcal{P}$ have no elements in common and the union of all of the subsets of A in $\mathcal{P}$ is A. This should not be confused with the power set of a set which is the set of all of the subsets of that set (including the empty set $\emptyset$). When a collection of sets satisfies condition (ii) from Definition 5.3.4 (no two distinct sets in the collection have any elements in common), the sets in the collection are said to be **pairwise disjoint**.

Example 5.3.5.

(i) *Consider the set $A = \{0, 1, 2, 3, 4, 5, 6, 7, 8, 9\}$.*

- $\mathcal{P}_1 = \{\{0, 2, 4, 6, 8\}, \{3, 5\}, \{7, 9\}\}$: *This is not a partition of A, as the union of the subsets of A in $\mathcal{P}_1$ does not equal A.*
- $\mathcal{P}_2 = \{\{0\}, \{2, 4\}, \{6, 8\}, \{0, 2, 3, 5, 7, 9\}\}$: *This is also not a partition of A, as the subsets of A in $\mathcal{P}_1$ are not pairwise disjoint (note that $\{2, 4\} \cap \{0, 2, 3, 5, 7, 9\} = \{2\} \neq \emptyset$).*
- $\mathcal{P}_3 = \{\{0, 3, 6\}, \{1, 2\}, \{4, 5, 8, 9\}, \{7\}\}$: *This is a partition of A.*

(ii) *Consider the set of integers $\mathbb{Z}$.*

- $\mathcal{P}_1 = \{\{\ldots, -3, -2, -1, 0\}, \{0\}, \{0, 1, 2, 3, \ldots\}\}$: *This is not a partition, as the subsets of $\mathbb{Z}$ in $\mathcal{P}_1$ are not pairwise disjoint.*
- *For each $n \in \mathbb{Z}$, such that $n \geq 0$, define the set $X_n = \{-n, n\}$ and let $\mathcal{P}_2 = \{X_n : n \in \mathbb{Z} \text{ for } n \geq 0\}$. Writing out the elements, we get that $\mathcal{P}_2 = \{\{0\}, \{-1, 1\}, \{-2, 2\}, \{-3, 3\}, \ldots\}$. This is a partition of $\mathbb{Z}$.*

Now we return to equivalence relations with the following definition.

Definition 5.3.6. *Let R be an equivalence relation of a set A, and let $a \in A$. The **equivalence class** of the element a, denoted by $[a]$, is the subset of A defined by*

$$[a] = \{x \in A : x \, R \, a\}.$$

Simply put, the equivalence class of the element $a \in A$ is the set of elements from A that are related to a.

Example 5.3.7. *Recall the set $A = \{1, 2, 3, 4, 5, 6\}$ and equivalence relation*

$$R = \{(1,1), (2,2), (2,3), (3,2), (3,3), (4,4), (4,5),$$
$$(4,6), (5,4), (5,5), (5,6), (6,4), (6,5), (6,6)\}$$

on set A mentioned earlier in this section. The equivalence class for each element in A is listed below.

$$[1] = \{a \in A : a \, R \, 1\} = \{1\}$$
$$[2] = \{a \in A : a \, R \, 2\} = \{2, 3\}$$
$$[3] = \{a \in A : a \, R \, 3\} = \{2, 3\}$$
$$[4] = \{a \in A : a \, R \, 4\} = \{4, 5, 6\}$$
$$[5] = \{a \in A : a \, R \, 5\} = \{4, 5, 6\}$$
$$[6] = \{a \in A : a \, R \, 6\} = \{4, 5, 6\}$$

There are three distinct equivalence classes $[1] = \{1\}, [2] = [3] = \{2, 3\}$, and $[4] = [5] = [6] = \{4, 5, 6\}$. First, you notice that the equivalence classes for 2 and 3 are equal and that the equivalence classes for $4, 5$, and 6 are also equal. This is not a coincidence as we shall soon see. The second observation is that the three equivalence classes form a partition for set A. This is also not by chance.

Before moving on to another example, we will address the first observation made in Example 5.3.7.

Theorem 5.3.8. *Let R be an equivalence relation on a set A. For elements $a, b \in A$, $[a] = [b]$ if and only if $a \, R \, b$.*

Proof. ($\Rightarrow$) Assume that $[a] = [b]$, and consider the element a. Since R is reflexive, we know that $a\,R\,a$. This implies by the definition of an equivalence class (Definition 5.3.6) that $a \in [a]$. Given that $[a] = [b]$, it follows that $a \in [b]$. This means that $a\,R\,b$.

($\Leftarrow$) Assume that $a\,R\,b$. To show $[a] \subseteq [b]$, let $x \in [a]$. It follows that $x\,R\,a$. But given that R is transitive, $x\,R\,a$ and $a\,R\,b$ implies that $x\,R\,b$. Consequently, we have $x \in [b]$ and $[a] \subseteq [b]$.

To show that $[b] \subseteq [a]$, let $y \in [b]$. It follows that $y\,R\,b$. Now given that $a\,R\,b$ and R is symmetric, it follows that $b\,R\,a$. Since R is transitive, $y\,R\,b$ and $b\,R\,a$ implies that $y\,R\,a$. Consequently, it follows that $y \in [a]$ and $[b] \subseteq [a]$. □

An immediate consequence of this result is the following corollary whose proof is an exercise.

Corollary 5.3.9. *Let R be an equivalence relation on a set A. For elements $a, b \in A$, either $[a] = [b]$ or $[a] \cap [b] = \emptyset$.*

Now our second example is given.

Example 5.3.10. *Recall that we proved in Proposition 5.3.2 that for $n \in \mathbb{Z}$ with $n \geq 1$, the relation R on the set $\mathbb{Z}$ defined by $a\,R\,b$ if and only if $a \equiv b \pmod{n}$ for $a, b \in \mathbb{Z}$ is an equivalence relation. The equivalence classes when $n = 5$ will now be determined. We will start with 0.*

$$[0] = \{x \in \mathbb{Z} : x\,R\,0\} = \{x \in \mathbb{Z} : x \equiv 0 \pmod{5}\}$$

For $x \equiv 0 \pmod 5$ to be satisfied, we must have $5|(x - 0)$ or $5|x$. Thus for x to be in $[0]$, it follows that $x = 5l$ for some integer l. Therefore, any multiple of 5 is in $[0]$ or

$$[0] = \{\dots, -20, -15, -10, -5, 0, 5, 10, 15, 20, \cdots\}.$$

We do not need to determine the equivalence class for any of the integers currently in $[0]$ as, by Theorem 5.3.8, they will determine the same equivalence class. To determine the next equivalence class, we will pick the integer 1 not in $[0]$.

$$[1] = \{x \in \mathbb{Z} : x\,R\,1\} = \{x \in \mathbb{Z} : x \equiv 1 \pmod{5}\}$$

For $x \equiv 1 \pmod 5$ to be satisfied, we must have $5|(x - 1)$. This implies $x - 1 = 5l$, for some integer l, yielding $x = 5l + 1$. Thus for x to be in

[1], *it will be of the form* $x = 5l + 1$, *for some integer* l, *or must have a remainder of 1 when divided by 5. This results in*

$$[1] = \{\ldots, -19, -14, -9, -4, 1, 6, 11, 16, 21, \cdots\}.$$

Again, we no longer need to determine the equivalence class of any integer in [1] *as it will determine the same equivalence class.*

 We move on to the integer 2, as it is not in the previous two equivalence classes.

$$[2] = \{x \in \mathbb{Z} : x \, R \, 2\} = \{x \in \mathbb{Z} : x \equiv 2 \pmod{5}\}$$

Since $5|(x - 2)$, *it follows that* $x - 2 = 5l$, *for some integer* l, *yielding* $x = 5l + 2$. *Thus this equivalence class contains all of the integers that have a remainder of 2 when divided by 5 or*

$$[2] = \{\ldots, -18, -13, -8, -3, 2, 7, 12, 17, 22, \cdots\}.$$

Continuing along the same lines, we obtain the final two equivalence classes

$$[3] = \{x \in \mathbb{Z} : x \, R \, 3\} = \{x \in \mathbb{Z} : x \equiv 3 \pmod{5}\}$$
$$= \{\ldots, -17, -12, -7, -2, 3, 8, 13, 18, 23, \cdots\}$$

and

$$[4] = \{x \in \mathbb{Z} : x \, R \, 4\} = \{x \in \mathbb{Z} : x \equiv 4 \pmod{5}\}$$
$$= \{\ldots, -16, -11, -6, -1, 4, 9, 14, 19, 24, \cdots\}.$$

Viewed in totality, the equivalence classes are listed here.

$$[0] = \{\ldots, -20, -15, -10, -5, 0, 5, 10, 15, 20, \cdots\}$$
$$[1] = \{\ldots, -19, -14, -9, -4, 1, 6, 11, 16, 21, \cdots\}$$
$$[2] = \{\ldots, -18, -13, -8, -3, 2, 7, 12, 17, 22, \cdots\}$$
$$[3] = \{\ldots, -17, -12, -7, -2, 3, 8, 13, 18, 23, \cdots\}$$
$$[4] = \{\ldots, -16, -11, -6, -1, 4, 9, 14, 19, 24, \cdots\}$$

These sets are nonempty. Given that any integer has a remainder of 0, 1, 2, 3, *or 4 when divided by 5, these sets are pairwise disjoint and their union is* $\mathbb{Z}$.

In general, given a positive integer n, the equivalence classes resulting from the equivalence relation on $\mathbb{Z}$, defined by $a \equiv b \pmod{n}$ for all $a, b \in \mathbb{Z}$, will be $[0], [1], [2], \ldots, [n-1]$.

In Examples 5.3.7 and 5.3.10, we noted that the equivalences classes formed a partition of the set on which the equivalence relation was defined. This is addressed in the following result.

Theorem 5.3.11. *Let A be a nonempty set. Any equivalence relation R on set A determines a partition of set A and any partition of A determines an equivalence relation on A.*

Proof. To prove the first part of this result, let R be an equivalence relation on A. Let $\mathcal{S} = \{[a] : a \in A\}$ be the set of all of the distinct equivalence classes determined by R. Let $x \in A$. Given that R is reflexive, $x\,R\,x$ and $x \in [x]$. Thus each equivalence class in $\mathcal{S}$ is nonempty. Now let $[x]$ and $[y]$ be two different equivalence classes in $\mathcal{S}$. We know by Corollary 5.3.9 that $[x] = [y]$ or $[x] \cap [y] = \emptyset$. Given that $[x]$ and $[y]$ are distinct, it follows that $[x] \cap [y] = \emptyset$.

To finish this section of the proof, we must show that the union of all of the equivalence classes in $\mathcal{S}$ is A or that $\bigcup_{[a]\in\mathcal{S}}[a] = A$. First note that each equivalence class $[a]$ is a subset of A. Thus it follows that $\bigcup_{[a]\in\mathcal{S}}[a] \subseteq A$. Now let x be an element in A. As we mentioned in the previous paragraph, $x \in [x]$ with $[x] \in \mathcal{S}$. This means $x \in \bigcup_{[a]\in\mathcal{S}}[a]$ and that $A \subseteq \bigcup_{[a]\in\mathcal{S}}[a]$.

For the second part of the theorem, let $\mathcal{P}$ be a partition of set A. We define a relation R on A by $a\,R\,b$ if and only if there is an element $X \in \mathcal{P}$ such that $a, b \in X$ for all $a, b \in A$. In other words, a and b are related if they are in the same subset of A in $\mathcal{P}$. Now let $x \in A$. Given that $\bigcup_{X\in\mathcal{P}} = A$, there exists a subset $Y \in \mathcal{P}$ such that $x \in Y$. This means $x\,R\,x$ and that R is reflexive. Now let $x, y \in A$ such that $x\,R\,y$. Thus there is a subset X of A in $\mathcal{P}$ such that $x, y \in X$. This implies that $y, x \in X$ and that $y\,R\,x$. Thus R is symmetric.

Finally, let $x, y, z \in A$ such that $x\,R\,y$ and $y\,R\,z$. This means there exists subsets X and Y of A in $\mathcal{P}$ such that $x, y \in X$ and $y, z \in Y$. Assume that $X \neq Y$ or that X and Y are different subsets of A in $\mathcal{P}$. But given that $y \in X$ and $y \in Y$, we have that $X \cap Y \neq \emptyset$. This is a contradiction as $\mathcal{P}$ is a partition of A. Consequently $X = Y$ and $x, z \in X$. Therefore, $x\,R\,z$ and R is transitive. $\qquad\square$

Exercises 5.3

(1) Let $A = \{a, b, c, d, e, f, g, h, i, j\}$.

   (a) Determine which of the following are a partition of set A.

      i. $\mathcal{P}_1 = \{\{a, e, i\}, \{b, d, f\}, \{c, g, h, j\}\}$
      ii. $\mathcal{P}_2 = \{\{a, d, g, j\}, \{b, e, f, h\}, \{c, g, i\}\}$
      iii. $\mathcal{P}_3 = \{\{a, b, c\}, \{d, e, f\}, \{g\}, \{h, i\}\}$

   (b) Determine the partition of A with the fewest number of subsets of A.

   (c) Determine the partition of A with the largest number of subsets.

(2) Consider the set of integers $\mathbb{Z}$. Determine which of the following are partitions of $\mathbb{Z}$.

   (a) $\mathcal{P}_1 = \{\{z \in \mathbb{Z} : z = 10q + r \text{ for } q, r \in \mathbb{Z} \text{ with } 1 \leq r \leq 9\}, \{z \in \mathbb{Z} : z = 10q \text{ for } q \in \mathbb{Z}\}\}$

   (b) $\mathcal{P}_2 = \{\{-n : n \in \mathbb{N}\}, \{n : n \in \mathbb{N}\}\}$

   (c) For $n \in \mathbb{N}$, let $X_n = \{-n, -n+1, \ldots, n-1, n\}$ and define $\mathcal{P}_3 = \{X_n : n \in \mathbb{N}\}\}$

   (d) $\mathcal{P}_4 = \{E, O\}$ where E is the set of even integers and O is the set of odd integers

(3) Consider the set of real numbers $\mathbb{R}$. Determine which of the following are a partition of $\mathbb{R}$.

   (a) $\mathcal{P}_1 = \{\{r \in \mathbb{R} : |r| \leq 1\}, \{r \in \mathbb{R} : |r| \geq 1\}\}$

   (b) $\mathcal{P}_2 = \{\{r \in \mathbb{R} : r \text{ is rational}\}, \{r \in \mathbb{R} : r \text{ is irrational}\}\}$

(4) Let $A = \{1, 2, 3, 4, 5\}$. Determine if the following relations on A are equivalence relations. If the relation is an equivalence relation, determine the list of equivalence classes. If it is not an equivalence relation, add the minimal number of ordered pairs until it becomes an equivalence relation.

   (a) $R_1 = \{(1, 1), (2, 2), (2, 3), (3, 2), (3, 3), (4, 3), (4, 4), (5, 5)\}$

   (b) $R_2 = \{(1, 1), (2, 2), (2, 3), (3, 2), (3, 3), (4, 4), (4, 5), (5, 4), (5, 5)\}$

   (c) $R_3 = \{(1, 1), (2, 2), (3, 3), (4, 4), (4, 5), (5, 3)\}$

(d) $R_4 = \{(1,1), (2,2), (3,3), (4,4), (5,5)\}$

(5) Determine if the following relations on the indicated sets are equivalence relations. Be sure to justify why the relation is or is not an equivalence relation.

(a) The relation R on set $\mathbb{Z}$ defined by $a\,R\,b$ if and only if $a+b$ is odd for all $a, b \in \mathbb{Z}$.

(b) The relation R on set $\mathbb{Z}$ defined by $a\,R\,b$ if and only if $2|(a+b)$ for all $a, b \in \mathbb{Z}$.

(c) The relation R on set $\mathbb{Z}$ defined by $a\,R\,b$ if and only if $a-b \geq 0$ for all $a, b \in \mathbb{Z}$.

(d) The relation R on set $\mathbb{Z}$ defined by $a\,R\,b$ if and only if $5a - 3b$ is even for all $a, b \in \mathbb{Z}$.

(e) The relation R on set $\mathbb{Z}$ defined by $a\,R\,b$ if and only if $5|(7a - 2b)$ for all $a, b \in \mathbb{Z}$.

(f) The relation R on set $\mathbb{R}$ defined by $x\,R\,y$ if and only if $x - y \in \mathbb{Q}$ for all $x, y \in \mathbb{R}$.

(g) The relation R on set $\mathbb{R}$ defined by $x\,R\,y$ if and only if $xy \geq 0$ for all $x, y \in \mathbb{R}$.

(h) The relation R on set $\mathbb{R}$ defined by $x\,R\,y$ if and only if $|x-y| < 1$ for all $x, y \in \mathbb{R}$.

(i) The relation R on set $\mathbb{R}$ defined by $x\,R\,y$ if and only if $xy \in \mathbb{Q}$ for all $x, y \in \mathbb{R}$.

(6) Show that each of the relations defined on the indicated sets are equivalence relations. Then determine the list of equivalence classes.

(a) The relation R on set $\mathbb{Z}$ defined by $a\,R\,b$ if and only if $a \equiv b$ (mod 7) for all $a, b \in \mathbb{Z}$.

(b) The relation R on set $\mathbb{Z}$ defined by $a\,R\,b$ if and only if $a+b$ is even for all $a, b \in \mathbb{Z}$.

(c) The relation R on set $\mathbb{Z}$ defined by $a\,R\,b$ if and only if $a^2 = b^2$ for all $a, b \in \mathbb{Z}$.

(d) The relation R on set $\mathbb{Z}$ defined by $a\,R\,b$ if and only if $3|(a+2b)$ for all $a, b \in \mathbb{Z}$.

(7) For the equivalence relation R on set $\mathbb{R}$ defined by $a\,R\,b$ if and only if $a^4 = b^4$ for all $a, b \in \mathbb{R}$, determine the equivalence class $[2]$.

(8) Show that the relation R on set $\mathbb{Q}$ defined by $\frac{a}{b}\,R\,\frac{c}{d}$ if and only if $ad = bc$ for $\frac{a}{b}, \frac{c}{d} \in \mathbb{Q}$ is an equivalence relation. Determine the equivalence classes $\left[\frac{1}{2}\right]$ and $\left[\frac{2}{5}\right]$.

(9) For the equivalence relation R on set $\mathbb{Q}$ defined by $a\,R\,b$ if and only if $a - b \in \mathbb{Z}$ for all $a, b \in \mathbb{Q}$, determine the equivalence class $\left[\frac{1}{5}\right]$.

(10) Prove Corollary 5.3.9: For an equivalence relation R on a set A, show that for elements $a, b \in A$, either $[a] = [b]$ or $[a] \cap [b] = \emptyset$.

CHAPTER 6

Functions

6.1 INTRODUCTION

As you know from your calculus and differential equations courses, functions play an important role in mathematics. You are familiar with using functions to model situations, with the goal of studying or manipulating the function to gain an insight into the original problem.

The main focus of Calculus I was the study of real-valued functions f where $f : \mathbb{R} \to \mathbb{R}$. Most were defined using an explicit formula such as $f(x) = x^2 + 1$ or

$$g(x) = \begin{cases} x + 5 & \text{if } x \leq 0 \\ \frac{1}{x} & \text{if } x > 0 \end{cases}.$$

We also graphed functions in the Cartesian plane to help analyze them. From an algebraic point of view, the graph of a real-valued function f is the collection of points $(x, f(x))$ in the plane. Therefore, a real-valued function, at its core, is a subset of the set $\mathbb{R} \times \mathbb{R}$ and thus a relation.

In this chapter, we will take a more general and global look at functions, using the language of relations to define and study them.

6.2 DEFINITION OF A FUNCTION

Definition 6.2.1. *Let A and B be sets. A **function** or **mapping** from A to B, denoted $f : A \to B$, is a relation from A to B such that every element $a \in A$ is related to exactly one element b in B.*

Given a function $f : A \to B$, it is a subset R of $A \times B$ such that for every element $a \in A$, there exists one and only one element $b \in B$ such that $(a, b) \in R$. This is denoted by $f(a) = b$ and we say that "f maps a to b" or "b is the image of a under f." In addition, when computing

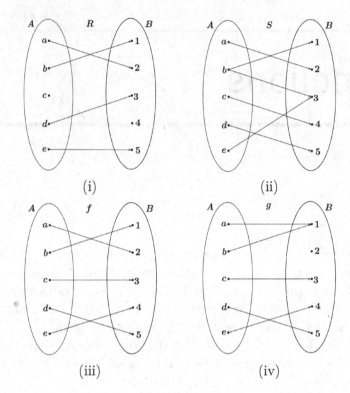

Figure 6.1 Relations from Example 6.2.2.

$f(a)$, we often say we are "plugging a into f" or "evaluating f at a." Every element A must be related to a single element in B, but not all elements in B need to have an element in A related to them. The set A is called the **domain** of f, denoted by $\mathrm{dom}(f) = A$, and the set B is called the **codomain** of f, denoted by $\mathrm{cod}(f) = B$.

Example 6.2.2. *Consider the set $A = \{a, b, c, d, e\}$ and $B = \{1, 2, 3, 4, 5\}$.*

(i) *Define the relation $R = \{(a, 2), (b, 1), (d, 3), (e, 5)\}$ from A to B, visualized in the top left in Figure 6.1. Since no element x in B exists such that $(c, x) \in R$, for $c \in A$, this relation is not a function.*

(ii) *Consider the relation $S = \{(a, 2), (b, 1), (b, 3), (c, 4), (d, 5), (e, 3)\}$ from A to B, visualized in the top right of Figure 6.1. Given that the element b in A is related to more than one element in B (both $(b, 1)$ and $(b, 3)$ are in S), this relation is not a function.*

(iii) *The relation $f = \{(a,2),(b,1),(c,3),(d,5),(e,4)\}$ from A to B, visualized in the bottom left in Figure 6.1, is a function. Every element in A is related to exactly one element in B. Another way to express this function is by saying $f(a) = 2, f(b) = 1, f(c) = 3, f(d) = 5,$ and $f(e) = 4$.*

(iv) *The relation $g = \{(a,1),(b,1),(c,3),(d,5),(e,4)\}$ from A to B, visualized in the bottom right in Figure 6.1, is also a function, even though a and b in A are related to the same element 1 in B and the element 2 in B has no element in A related to it.*

Example 6.2.3.

(i) *Consider the relation R defined on $\mathbb{Z}$ by $n\,R\,m$ if and only if $|n| = |m|+1$, for all $n, m, \in \mathbb{Z}$. Given that $|2| = |1|+1$ and $|2| = |-1|+1$, it has $2\,R\,1$ and $2\,R-1$, implying that relation R is not a function.*

(ii) *The relation S defined on $\mathbb{Z}$ by $n\,S\,m$ if and only if $|n| = m+1$, for all $n, m \in \mathbb{Z}$, is a function. Since $m = |n| - 1$, each integer $n \in \mathbb{Z}$ is mapped to exactly one element $m \in \mathbb{Z}$.*

(iii) *Explicit formulas can be used to define a function. As an example, consider the function $f : \mathbb{R} \times \mathbb{R} \to \mathbb{R}$ defined by $f(x,y) = x^2 - xy + y^3$, for all $(x,y) \in \mathbb{R} \times \mathbb{R}$.*

Definition 6.2.4. *Let $f : A \to B$ be a function, and let $X \subseteq A$. The **image** X under f is the subset $f(X)$ of B where*

$$f(X) = \{b \in B : f(a) = b \text{ for some } a \in X\}.$$

Another way to view the image of X under f is $f(X) = \{b \in B : (a,b) \in f \text{ for some } a \in X\}$ or $f(X) = \{f(a) : a \in X\}$.

Example 6.2.5.

(i) *Recall function $f : \{a,b,c,d,e\} \to \{1,2,3,4,5\}$ defined by*

$$f = \{(a,2),(b,1),(c,3),(d,5),(e,4)\}$$

from Example 6.2.2 (iii). For this function, we have $f(\{a,c,d\}) = \{2,3,5\}$, $f(\{b\}) = \{1\}$, and $f(\{a,b,c,d,e\}) = \{1,2,3,4,5\}$.

(ii) *Define a function $h : \mathbb{Z} \to \mathbb{Z}$ by $h(n) = |n| + 1$ for $n \in \mathbb{Z}$. Here we get that*

$$h(\{-3, -2, -1, 0, 1, 2, 3\}) = \{1, 2, 3, 4\}$$

and

$$h(\mathbb{Z}) = \{1, 2, 3, 4, 5, \ldots\} = \mathbb{N}.$$

Computing $f(\{a, b, c, d, e\})$ in Example 6.2.5 (i) and $h(\mathbb{Z})$ in Example 6.2.5 (ii) motivates the following definition.

Definition 6.2.6. *Let $f : A \to B$ be a function. The **range** of f, denoted $\operatorname{ran}(f)$, is the image $f(A)$.*

For function f from Example 6.2.2 (iii), $f(A) = B$, meaning that $\operatorname{ran}(f) = B$. For function g from Example 6.2.2 (iv), $g(A) = \{1, 3, 4, 5\}$, meaning that $\operatorname{ran}(g) = \{1, 3, 4, 5\}$.

Example 6.2.7. *Consider the following examples.*

(i) *Let $A = \{-2, -1, 0, 1, 2\}$ and $B = \{0, 1, 2, 3, 4\}$. Define the function $f : A \to B$ by $f(a) = a^2$ for all $a \in A$. For this function, $\operatorname{ran}(f) = \{0, 1, 4\}$.*

(ii) *Define the function $h : \mathbb{Q} \to \mathbb{Z}$ by $h\left(\frac{n}{m}\right) = n + m$ for $\frac{n}{m} \in \mathbb{Q}$. To determine the range, we need to determine the integers $z \in \mathbb{Z}$ such that $h\left(\frac{n}{m}\right) = z$ or $n + m = z$ for some $\frac{n}{m} \in \mathbb{Q}$. Given that $z = z + 0 = z - 1 + 1 = (z - 1) + 1$ with $1 \neq 0$, we have that $\frac{z-1}{1} \in \mathbb{Q}$ and $h\left(\frac{z-1}{1}\right) = z - 1 + 1 = z$. Thus we have $\operatorname{ran}(h) = \mathbb{Z}$.*

(iii) *Consider the real-valued function defined by $j(x) = \sqrt{x^2 - 2x - 3}$. Since $j(0)$ is undefined, the domain is not $\mathbb{R}$. Solving $x^2 - 2x - 3 \geq 0$ results in $x \leq -1$ or $x \geq 3$. Thus the $\operatorname{dom}(j) = \{x \in \mathbb{R} : x \leq -1 \text{ or } x \geq 3\} = (-\infty, -1] \cup [3, \infty)$.*

Given that $\sqrt{x^2 - 2x - 3} \geq 0$ for all values of x in the domain, it follows from calculus that the range of j is $\{y \in \mathbb{R} : y \geq 0\}$ or the interval $[0, \infty)$.

Before we state and prove some properties of functions, one important type of function needs to be introduced.

Definition 6.2.8. *Let A be a set. The **identity function** on A, denoted i_A, is the function defined by $i_A(a) = a$ for all $a \in A$.*

Simply stated, the identity function on a set maps every element in the set to itself.

Proposition 6.2.9. *Let* $f : A \to B$ *be a function, and let* X_1 *and* X_2 *be subsets of* A.

(i) *If* $X_1 \subseteq X_2$, *then* $f(X_1) \subseteq f(X_2)$.

(ii) $f(X_1 \cup X_2) = f(X_1) \cup f(X_2)$.

(iii) $f(X_1 \cap X_2) \subseteq f(X_1) \cap f(X_2)$.

Only item (ii) from Proposition 6.2.9 will be proven here. The remaining two results are exercises.

Proof of Proposition 6.2.9 (ii). To show that $f(X_1 \cup X_2) \subseteq f(X_1) \cup f(X_2)$, let $b \in f(X_1 \cup X_2)$. Thus there exists an element $a \in X_1 \cup X_2$ such that $f(a) = b$. Since $a \in X_1 \cup X_2$, we have that $a \in X_1$ or $a \in X_2$. If $a \in X_1$, then $b = f(a) \in f(X_1)$ and $b \in f(X_1) \cup f(X_2)$. If $a \in X_2$, then $b = f(a) \in f(X_2)$ and $b \in f(X_1) \cup f(X_2)$. Either way, it follows that $b \in f(X_1) \cup f(X_2)$ and $f(X_1 \cup X_2) \subseteq f(X_1) \cup f(X_2)$.

To show that $f(X_1) \cup f(X_2) \subseteq f(X_1 \cup X_2)$, let $b \in f(X_1) \cup f(X_2)$. Then $b \in f(X_1)$ or $b \in f(X_2)$. If $b \in f(X_1)$, there exists an element $a \in X_1$ such that $f(a) = b$. Since $a \in X_1$, we know that $a \in X_1 \cup X_2$, implying $b = f(a) \in f(X_1 \cup X_2)$. If $b \in f(X_2)$, there exists an element $c \in X_2$ such that $f(c) = b$. Since $c \in X_2$, we know that $c \in X_1 \cup X_2$, implying $b = f(c) \in f(X_1 \cup X_2)$. This results in $f(X_1) \cup f(X_2) \subseteq f(X_1 \cup X_2)$. $\square$

Definition 6.2.10. *Let* $f : A \to B$ *be a function, and let* Y *be a subset of* B. *The* **preimage** *of* Y *under* f, *denoted* $f^{-1}(Y)$ *is the subset of* A *determined by* $f^{-1}(Y) = \{a \in A : f(a) \in Y\}$.

In other words, an element a in A is in $f^{-1}(Y)$ if and only if $f(a)$ is in Y.

Example 6.2.11.

(i) *Recall the function* $f : \{-2, -1, 0, 1, 2\} \to \{0, 1, 2, 3, 4\}$ *defined by* $f(a) = a^2$ *for all* $a \in A$. *Here* $f^{-1}(\{0, 1\}) = \{-1, 0, 1\}$ *and* $f^{-1}(\{4\}) = \{-2, 2\}$.

(ii) *Let $f : \mathbb{Z} \times \mathbb{Z} \to \mathbb{N}$ be a function defined by $f(a, b) = a^2 + b^2$ for all $(a, b) \in \mathbb{Z} \times \mathbb{Z}$. For this function,*

$$f^{-1}(\{1, 2, 3, 4\}) = \{(0, -1), (0, 1), (-1, 0), (1, 0), (-1, -1), (-1, 1),$$
$$(1, -1), (1, 1), (-2, 0), (2, 0), (0, -2), (0, 2)\}$$

and

$$f^{-1}(\{5\}) = \{(-1, -2), (-1, -2), (1, -2), (1, 2)\}.$$

However, $f^{-1}(\{7\}) = \emptyset$, as 7 can't be expressed as the sum of the squares of two integers.

(iii) *Let $g : \mathbb{R} \to \mathbb{R}$ be a function defined by $g(x) = \frac{x}{2x-4}$ for all $x \in \mathbb{R}$. To compute $g^{-1}((-\infty, 0])$, we need to find the values of $x \in \mathbb{R}$ such that $g(x) \le 0$ or $\frac{x}{2x-4} \le 0$. The first case is $x \le 0$ and $2x - 4 > 0$, which is not possible. The second case is $x \ge 0$ and $2x - 4 < 0$, resulting in $0 \le x < 2$. Thus we have $g^{-1}((-\infty, 0]) = [0, 2)$.*

Proposition 6.2.12. *Let $f : A \to B$ be a function, and let Y_1 and Y_2 be subsets of B.*

(i) *If $Y_1 \subseteq Y_2$, then $f^{-1}(Y_1) \subseteq f^{-1}(Y_2)$.*

(ii) *$f^{-1}(Y_1 \cup Y_2) = f^{-1}(Y_1) \cup f^{-1}(Y_2)$.*

(iii) *$f^{-1}(Y_1 \cap Y_2) = f^{-1}(Y_1) \cap f^{-1}(Y_2)$.*

Only item (iii) from Proposition 6.2.12 will be proven here. The proofs of the remaining two results are exercises.

Proof of Proposition 6.2.12 (iii). To show that show $f^{-1}(Y_1 \cap Y_2) \subseteq f^{-1}(Y_1) \cap f^{-1}(Y_2)$, let $a \in f^{-1}(Y_1 \cap Y_2)$. This implies that $f(a) \in Y_1 \cap Y_2$. Thus we have $f(a) \in Y_1$ and $f(a) \in Y_2$. Since $f(a) \in Y_1$, it follows that $a \in f^{-1}(Y_1)$. Similarly, the fact that $f(a) \in Y_2$ yields $a \in f^{-1}(Y_2)$. This results in $a \in f^{-1}(Y_1) \cap f^{-1}(Y_2)$.

To show that $f^{-1}(Y_1) \cap f^{-1}(Y_2) \subseteq f^{-1}(Y_1 \cap Y_2)$, let $a \in f^{-1}(Y_1) \cap f^{-1}(Y_2)$. Then we have $a \in f^{-1}(Y_1)$ and $a \in f^{-1}(Y_2)$. Since $a \in f^{-1}(Y_1)$, we know that $f(a) \in Y_1$. Similarly, since $a \in f^{-1}(Y_2)$, it follows that $f(a) \in Y_2$. This yields $f(a) \in Y_1 \cap Y_2$, which results in $a \in f^{-1}(Y_1 \cap Y_2)$. □

Now we introduce the last concept of this section.

Definition 6.2.13. *Let $f : A \to B$ and $g : A \to C$ be functions. These two functions are **equal**, denoted by $f = g$, if and only if $f(a) = g(a)$ for all $a \in A$.*

Note that for two functions to be equal, they must have the same domain. However, they do not need to have the same codomain as long as they have identical function values for the same input values from the domain. For example, consider the functions $f : \mathbb{Z} \to \mathbb{Z}$, defined by $f(n) = n^2 + 2n + 2$, and the function $g : \mathbb{Z} \to \mathbb{N}$, defined by $g(n) = (n+1)^2 + 1$, are equal. They have the same domain and equal function values. However, note that the $\text{cod}(f) = \mathbb{Z}$ while $\text{cod}(g) = \mathbb{N}$.

Exercises 6.2

(1) Determine if the indicated relation on the given sets is a function. Be sure to justify your answer.

    (a) Let $A = \{0, 2, 4, 6, 8\}$ and $B = \{a, b, c, d, e\}$.
       i. $R_1 = \{(0, a), (2, e), (4, d), (6, c), (8, c)\}$
      ii. $R_2 = \{(0, b), (2, a), (4, e), (8, d)\}$
     iii. $R_3 = \{(0, e), (2, d), (4, c), (4, a), (6, e), (8, b)\}$

    (b) The relation R defined on $\mathbb{N}$ by $n \, R \, m$ if and only if $n + m = 10$ for all $n, m \in \mathbb{N}$.

    (c) The relation R defined on $\mathbb{N}$ by $n \, R \, m$ if and only if $2n - m = 1$ for all $n, m \in \mathbb{N}$.

    (d) The relation R defined from $\mathbb{Z}$ to $\mathbb{N}$ by $n \, R \, m$ if and only if $|n| \le m$ for all $n \in \mathbb{Z}$ and $m \in \mathbb{N}$.

    (e) The relation R defined from $\mathbb{Z}$ to $\mathbb{N}$ by $n \, R \, m$ if and only if $n^2 + 1 = m$ for all $n \in \mathbb{Z}$ and $m \in \mathbb{N}$.

    (f) The relation R defined on $\mathbb{Z}$ by $a \, R \, b$ if and only if $a = \frac{b}{2}$ for all $a, b \in \mathbb{Z}$.

    (g) The relation R defined from $\mathbb{Q}$ to $\mathbb{Z}$ by $\frac{a}{b} \, R \, n$ if and only if $ab = n$ for all $\frac{a}{b} \in \mathbb{Q}$ and $n \in \mathbb{Z}$.

    (h) The relation R defined from $\mathbb{R}$ to $\mathbb{R}^+$ by $x \, R \, y$ if and only if $2^x = y$ for all $x \in \mathbb{R}$ and $y \in \mathbb{R}^+$.

    (i) The relation R defined on $\mathbb{R}$ by $x \, R \, y$ if and only if $x^2 + y^2 = 1$ for all $x, y \in \mathbb{R}$.

    (j) The relation R defined on $\mathbb{R}$ by $x \, R \, y$ if and only if $(|x| + 1)y = 1$ for all $x, y \in \mathbb{R}$.

(2) For each of the functions $f_i : A_i \to \mathbb{R}$ listed below, determine the largest domain A_i.

(a) $f_1(x) = \frac{x+1}{x}$.

(c) $f_3(x) = \sqrt{1 - x^2}$.

(b) $f_2(x) = \sqrt{5x + 1}$.

(d) $f_4(x) = \frac{1}{x^2+3x-10}$.

(3) Find all of the functions from $\{0, 1\}$ to $\{0, 1\}$.

(4) Let $A = \{1, 2\}$ and $B = \{a, b, c\}$. Does there exist a function $f : A \to B$ such that $\operatorname{ran}(f) = B$? Justify your answer.

(5) Determine the indicated sets for each of the following functions with E denoting the even integers and O denoting the odd integers.

(a) Function $f : A \to B$, with $A = \{0, 2, 4, 6, 8\}$ and $B = \{a, b, c, d, e\}$, defined by $f = \{(0, b), (2, c), (4, d), (6, c), (8, b)\}$.

    i. $f(\{0, 4, 8\})$ iii. $f^{-1}(\{b, e\})$

    ii. $\operatorname{ran}(f)$

(b) Function $f : \mathbb{N} \to \mathbb{N}$ defined by $f(n) = 2n + 1$ for all $n \in \mathbb{N}$.

    i. $f(\{-2, -1, 0, 1, 2\})$ iii. $f^{-1}(\{13\})$

    ii. $\operatorname{ran}(f)$ iv. $f^{-1}(E)$

(c) Function $f : \mathbb{Z} \to \mathbb{Z}$ defined by $f(n) = 3n + 2$ for all $n \in \mathbb{N}$.

    i. $f(\{-2, -1, 0, 1, 2\})$ ii. $\operatorname{ran}(f)$

(d) Function $f : \mathbb{Z} \to \mathbb{N}$ defined by $f(n) = |n| + 1$ for all $n \in \mathbb{Z}$.

    i. $f(O)$ iii. $f^{-1}(\{1, 2, 3, 4, 5\})$

    ii. $\operatorname{ran}(f)$ iv. $f^{-1}(\{m \in \mathbb{N} : m \geq 10\})$

(e) Function $f : \mathbb{Z} \to \mathbb{Z}$ defined by $f(n) = n \pmod 7$ for all $n \in \mathbb{Z}$.

    i. $f(\{7m : m \in \mathbb{Z}\})$ iii. $f^{-1}(\{3\})$

    ii. $\operatorname{ran}(f)$ iv. $f^{-1}(\{5\})$

(f) Function $f : \mathbb{Z} \to \mathbb{Z}$ defined by $f(n) = (-2)^n$ for all $n \in \mathbb{Z}$.

i. $f(E)$

ii. $\operatorname{ran}(f)$

iii. $f^{-1}(\{-2, 4\})$

iv. $f^{-1}\left(\{-\frac{1}{8}, -\frac{1}{2}, -2, -8\}\right)$

(g) Function $f : \mathbb{R} \to \mathbb{R}$ defined by $f(x) = x^2 + 1$ for all $x \in \mathbb{R}$.

i. $f([-2, 2])$

ii. $\operatorname{ran}(f)$

iii. $f^{-1}([9, \infty))$

(h) Function $f : \mathbb{R} \to \mathbb{R}$ defined by $f(x) = 2\sin(x) - 1$ for all $x \in \mathbb{R}$.

i. $\operatorname{ran}(f)$

ii. $f^{-1}(\{0\})$

(i) For the function $f : \mathbb{R} \to \mathbb{R}$ defined by $f(x) = \frac{3x}{x^2+1}$ for all $x \in \mathbb{R}$, find $f^{-1}([0, \infty))$.

(6) Let $f : A \to B$ be a function. Prove that if X is a nonempty subset of A that $f(X) \neq \emptyset$.

(7) Let $f : A \to B$ be a function. Prove that $f(\emptyset) = \emptyset$.

(8) Prove Proposition 6.2.9 (i): for function $f : A \to B$ with X_1 and X_2 subsets of A, prove that if $X_1 \subseteq X_2$, then $f(X_1) \subseteq f(X_2)$.

(9) Prove Proposition 6.2.9 (iii): for function $f : A \to B$ with X_1 and X_2 subsets of A, prove that $f(X_1 \cap X_2) \subseteq f(X_1) \cap f(X_2)$.

(10) Find an example of a function $f : A \to B$ with $X_1, X_2 \subseteq A$ such that $f(X_1 \cap X_2) \neq f(X_1) \cap f(X_2)$.

(11) For function $f : A \to B$ with $X_1, X_2, \ldots, X_n$ subsets of A, for $n \in \mathbb{N}$, prove that $f(X_1 \cup X_2 \cup \cdots \cup X_n) = f(X_1) \cup f(X_2) \cup \cdots \cup f(X_n)$.

(12) For function $f : A \to B$ with $Y_1, Y_2, \ldots, Y_n$ subsets of B, for $n \in \mathbb{N}$, prove that $f^{-1}(Y_1 \cap Y_2 \cap \cdots \cap Y_n) = f^{-1}(Y_1) \cap f^{-1}(Y_2) \cap \cdots \cap f^{-1}(Y_n)$.

(13) For function $f : A \to B$ with X_1 and X_2 subsets of A, prove that $f(X_1) - f(X_2) \subseteq f(X_1 - X_2)$.

(14) Given an example of a function $f : A \to B$ showing that it is not necessarily true that $f(X_1 - X_2) \subseteq f(X_1) - f(X_2)$ for subsets X_1 and X_2 of A.

(15) Prove Proposition 6.2.12 (i): for function $f : A \to B$ with Y_1 and Y_2 subsets of B such that $Y_1 \subseteq Y_2$, prove that $f^{-1}(Y_1) \subseteq f^{-1}(Y_2)$.

(16) Prove Proposition 6.2.12 (ii): for function $f : A \to B$ with Y_1 and Y_2 subsets of B, prove that $f^{-1}(Y_1 \cup Y_2) = f^{-1}(Y_1) \cup f^{-1}(Y_2)$.

(17) Let $f : A \to B$ be a function, and let Y_1 and Y_2 be subsets of B. Prove that $f^{-1}(Y_1 - Y_2) = f^{-1}(Y_1) - f^{-1}(Y_2)$.

(18) Let $f : A \to B$ be a function, let X be a subset of A, and let Y be a subset of B.

   (a) Prove $X \subseteq f^{-1}(f(X))$.   (b) Prove $f(f^{-1}(Y)) \subseteq Y$.

(19) Let $f : \mathbb{R} \to \mathbb{R}$ be a function defined by $f(x) = ax + 3$, for $a \in \mathbb{R}$, and let $g : \mathbb{R} \to \mathbb{R}$ be a function defined by $g(x) = \frac{x^3 + bx^2 + x + 3}{x^2 + 1}$, for $b \in \mathbb{R}$. Determine values of a and b such that $f = g$.

6.3 ONE-TO-ONE AND ONTO FUNCTIONS

We start this section with an definition.

Definition 6.3.1. *Let $f : A \to B$ be a function.*

(i) *The function f is **one-to-one** or **injective** if and only if for all $a_1, a_2 \in A$, $a_1 \neq a_2$ implies $f(a_1) \neq f(a_2)$.*

(ii) *The function f is **onto** or **surjective** if and only if for every element $b \in B$, there exists an element $a \in A$ such that $f(a) = b$.*

(iii) *The function f is **bijective** or **a bijection** if and only if it is both one-to-one (injective) and onto (surjective).*

Example 6.3.2. *Consider the following functions.*

(i) *The function $f_1 : \{a, b, c, d\} \to \{1, 2, 3, 4\}$, defined by $f_1(a) = 1, f_1(b) = 1, f_1(c) = 4$, and $f_1(d) = 3$, is shown in Figure 6.2 (i). It is not one-to-one as $a \neq b$ with $f_1(a) = f_1(b) = 1$, and it is not onto as no element in A is mapped to the element $2 \in B$.*

*Note: As motivated by the visualization of function f_1 in Figure 6.2 (i), when a function maps two distinct elements to the same image, it is sometimes called a **collision**.*

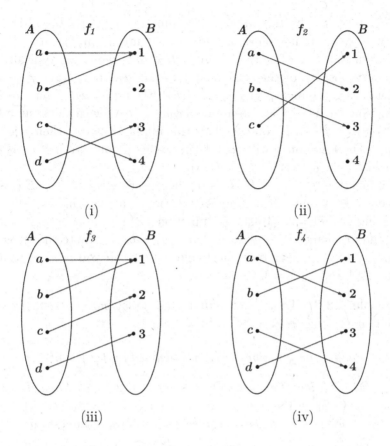

(i)

(ii)

(iii)

(iv)

Figure 6.2 Functions from Example 6.3.2.

(ii) *The function f_2 : $\{a, b, c\} \rightarrow \{1, 2, 3, 4\}$, defined by $f_2(a) = 2, f_2(b) = 3$, and $f_2(c) = 1$, is shown in Figure 6.2 (ii). This function is one-to-one, but is not onto given that no element in A is mapped to the element 4 in B.*

(iii) *The function f_3 : $\{a, b, c, d\} \rightarrow \{1, 2, 3\}$, defined by $f_3(a) = 1, f_3(b) = 1, f_3(c) = 2$, and $f_3(d) = 3$, is shown in Figure 6.2 (iii). This function is not one-to-one since $a \neq b$ with $f_3(a) = f_3(b) = 1$. However, it is onto.*

Note: For this function, we have $\text{ran}(f_3) = B$.

(iv) *The function f_4 : $\{a, b, c, d\} \rightarrow \{1, 2, 3, 4\}$, defined by $f_4(a) = 2, f_4(b) = 1, f_4(c) = 4$, and $f_4(d) = 3$, is shown in Figure 6.2 (iv). Since this function is one-to-one and onto, it is a bijection.*

As motivated by Example 6.3.2, a function $f : A \rightarrow B$ will be one-to-one if there are no collisions and will be onto if $\text{ran}(f) = B$.

When showing a function $f : A \rightarrow B$ is one-to-one, we typically use the contrapositive of the statement "If $a_1 \neq a_2$, then $f(a_1) \neq f(a_2)$," which is the statement, "If $f(a_1) = f(a_2)$, then $a_1 = a_2$." Thus to show a function $f : A \rightarrow B$ is one-to-one, let $a_1, a_2 \in A$ with the assumption that $f(a_1) = f(a_2)$. If you can show this implies $a_1 = a_2$, then f is one-to-one. The function f will not be one-to-one if you can find $a_1, a_2 \in A$ such that $a_1 \neq a_2$, yet $f(a_1) = f(a_2)$.

A function $f : A \rightarrow B$ is onto when every element $b \in B$ has an element $a \in A$ that maps to it. As mentioned above, this is equivalent to saying $\text{ran}(f) = B$. To show a function $f : A \rightarrow B$ is onto, let $b \in B$ and find an element $a \in A$ such that $f(a) = b$. Now note that it is not necessary to document how you found $a \in A$. All you need to show is that $a \in A$ and $f(a) = b$.

Example 6.3.3. *We will determine if the following functions are one-to-one, onto, or bijective.*

(i) *Consider the function* $f : \mathbb{Z} \rightarrow \mathbb{Z}$ *defined by* $f(n) = n^2 + 1$.

   (a) *Consider the two elements* -2 *and* 2 *in* $\mathbb{Z}$. *Since* $f(-2) = (-2)^2 + 1 = 5$ *and* $f(2) = 2^2 + 1 = 5$, *we have* $f(-2) = f(2)$ *with* $-2 \neq 2$. *This means function* f *is not one-to-one.*

   (b) *For any* $n \in \mathbb{Z}$, *we have* $n^2 \geq 0$ *implying that* $n^2 + 1 \geq 0 + 1 = 1$. *Thus* $f(n) \geq 1$ *for all* $n \in \mathbb{Z}$. *However,* -3 *is an integer with* $-3 < 1$. *Thus there is no element* $n \in \mathbb{Z}$ *such that* $f(n) = -3$. *This means that* f *is not onto.*

(ii) *Consider the function* $g : \mathbb{Z} \times \mathbb{Z} \rightarrow \mathbb{Z}$ *defined by* $f(n, m) = n + m + 1$.

   (a) *Consider the two elements* $(-2, 2)$ *and* $(-3, 3)$ *in* $\mathbb{Z} \times \mathbb{Z}$. *Given that* $g(-2, 2) = -2 + 2 + 1 = 1$ *and* $g(-3, 3) = -3 + 3 + 1 = 1$, *we have* $g(-2, 2) = g(-3, 3)$ *with* $(-2, 2) \neq (-3, 3)$. *Thus function* g *is not one-to-one.*

   (b) *Let* $b \in \mathbb{Z}$. *Rewriting, we have* $b = b + 0 = b - 1 + 1 = b + (-1) + 1$, *with* b *and* -1 *integers. Thus consider the element* $(b, -1) \in \mathbb{Z} \times \mathbb{Z}$. *It then follows that* $g(b, -1) = b + (-1) + 1 = b$ *and that* g *is onto.*

(iii) *Consider the function* $h : \mathbb{Z} \rightarrow \mathbb{Z}$ *defined by* $h(n) = 7n + 3$.

(a) Let $n_1, n_2 \in \mathbb{Z}$ and assume that $h(n_1) = h(n_2)$. This implies that $7n_1 + 3 = 7n_2 + 3$. The calculation

$$7n_1 + 3 = 7n_2 + 3$$
$$7n_1 = 7n_2$$
$$n_1 = n_2$$

shows $h(n_1) = h(n_2)$ implies $n_1 = n_2$. Thus h is one-to-one.

(b) Given that $h(n) = 7n+3$ suggests that the only integers in the range of h are those that have a remainder of 3 when divided by 7. Consider the integer 8. If there did exist an $n \in \mathbb{Z}$ such that $h(n) = 8$, then $7n + 3 = 8$ or $7n = 5$. Since 5 is not a multiple of 7, there is no integer $n \in \mathbb{Z}$ such that $f(n) = 8$. This means that h is not onto.

(iv) Consider the function $f : \mathbb{R} - \{2\} \to \mathbb{R} - \{3\}$ defined by $f(x) = \frac{3x}{x-2}$.

(a) Let $x_1, x_2 \in \mathbb{R} - \{2\}$ and assume that $f(x_1) = f(x_2)$. This implies that $\frac{3x_1}{x_1 - 2} = \frac{3x_2}{x_2 - 2}$. The calculation

$$\frac{3x_1}{x_1 - 2} = \frac{3x_2}{x_2 - 2}$$
$$3x_1(x_2 - 2) = 3x_2(x_1 - 2)$$
$$3x_1 x_2 - 6x_1 = 3x_1 x_2 - 6x_2$$
$$-6x_1 = -6x_2$$
$$x_1 = x_2.$$

shows $f(x_1) = f(x_2)$ implies $x_1 = x_2$. Thus f is one-to-one.

(b) To show f is onto, start by letting $y \in \mathbb{R} - \{3\}$. We must show there exists an $x \in \mathbb{R} - \{2\}$ such that $f(x) = y$ or that $\frac{3x}{x-2} = y$. To see if such an x exists, solve the equation for x.

$$\frac{3x}{x - 2} = y$$
$$3x = y(x - 2)$$
$$3x = xy - 2y$$
$$3x - xy = -2y$$
$$x(3 - y) = -2y$$
$$x = \frac{-2y}{3 - y}$$

Now x is defined and a real number as $y \neq 3$. Furthermore, we have that $x \neq 2$. If it did, then $\frac{-2y}{3-y} = 2$, implying that $-2y = 2(3 - y)$ or that $-2y = 6 - 2y$. This means $0 = 6$, a contradiction. It follows that $x \in \mathbb{R} - \{2\}$. However, we are not done. We must show that $f(x) = y$, which is demonstrated by the following calculation:

$$f(x) = f\left(\frac{-2y}{3-y}\right) = \frac{3 \cdot \frac{-2y}{3-y}}{\frac{-2y}{3-y} - 2} = \frac{\frac{-6y}{3-y}}{\frac{-2y}{3-y} - \frac{6-2y}{3-y}}$$

$$= \frac{\frac{-6y}{3-y}}{\frac{-6}{3-y}} = \frac{-6y}{3-y} \cdot \frac{3-y}{-6} = y$$

Since $f(x) = y$, the function f is onto.

(c) Given that f is both one-to-one and onto, it is a bijection.

We end this section with a simple lemma.

Lemma 6.3.4. *Let* $f : A \to B$ *be an one-to-one function. Then there exists a bijection from A to $f(A)$.*

Proof. Define the function $g : A \to f(A)$ by $g(a) = f(a)$ for all $a \in A$. To show that g is one-to-one, let $a, b \in A$ such that $g(a) = g(b)$. This implies that $f(a) = f(b)$. Given that f is one-to-one, this means $a = b$ and that g is one-to-one.

To show that g is onto, let $y \in f(A)$. By definition, there exists an element $x \in A$ such that $f(x) = y$. Thus by the definition of g, it follows that $g(x) = f(x) = y$ and g is onto. Since g is both one-to-one and onto, it is a bijection. $\qquad\square$

Exercises 6.3

(1) Determine if the indicated function defined on the given sets is one-to-one, onto, or a bijection. Be sure to justify your answer.

    (a) Let $A = \{0, 2, 4, 6, 8\}$ and $B = \{a, b, c, d, e\}$.

        i. $f_1 = \{(0, a), (2, e), (4, d), (6, c), (8, c)\}$

        ii. $f_2 = \{(0, b), (2, a), (4, e), (6, b), (8, d)\}$

    (b) Function $f : \mathbb{Z} \to \mathbb{Z}$ defined by $f(n) = 3n + 1$ for all $n \in \mathbb{Z}$.

    (c) Function $f : \mathbb{Z} \to \mathbb{Z}$ defined by $f(n) = 1 - n$ for all $n \in \mathbb{Z}$.

(d) Function $f : \mathbb{Z} \to \mathbb{N}$ defined by $f(n) = |n| + 1$ for all $n \in \mathbb{Z}$.

(e) Function $f : \mathbb{Z} \times \mathbb{Z} \to \mathbb{Z}$ defined by $f(n, m) = 3n + m$ for all $(n, m) \in \mathbb{Z} \times \mathbb{Z}$.

(f) Function $f : \mathbb{Z} \to \mathbb{Z} \times \mathbb{Z}$ defined by $f(n) = (n - 1, n + 1)$ for all $n \in \mathbb{Z}$.

(g) Function $f : \mathbb{R} \to \mathbb{R}$ defined by $f(x) = 7x^3 + 1$ for all $x \in \mathbb{R}$.

(h) Function $f : \mathbb{R} \to \mathbb{R}$ defined by $f(x) = \sqrt{x^2 + 1}$ for all $x \in \mathbb{R}$.

(i) Function $f : \mathbb{R} - \{1\} \to \mathbb{R} - \{1\}$ defined by $f(x) = \frac{x}{x+1}$ for all $x \in \mathbb{R} - \{1\}$.

(j) Function $f : \mathbb{R} - \{-3\} \to \mathbb{R} - \{7\}$ defined by $f(x) = \frac{7x}{x+3}$ for all $x \in \mathbb{R} - \{3\}$.

(k) Function $f : \mathbb{R} \to \mathbb{R}$ defined by $f(x) = 3^x$ for all $x \in \mathbb{R}$.

(l) Function $f : \mathbb{R} \to \mathbb{R}^+ \cup \{0\}$ defined by $f(x) = (x - 1)^2$ for all $x \in \mathbb{R}$.

(2) Let $f : A \to B$ be a function. Prove that f is a bijection if and only if for every $b \in B$ there exists a unique $a \in A$ such that $f(a) = b$.

(3) Let $f : A \to B$ be a function. Prove that f is a bijection if and only if for every $b \in B$, $f^{-1}(\{b\}) = \{a\}$ for a unique $a \in A$.

(4) Let $f : A \to B$ be a function, and let X be a subset of A. Prove that if f is one-to-one, then $f^{-1}(f(X)) = X$.

(5) Let $f : A \to B$ be a function, and let Y be a subset of B. Prove that if f is onto, then $f(f^{-1}(Y)) = Y$.

(6) Given a nonempty set A, prove there must always exist a bijection from A to A.

6.4   COMPOSITION OF FUNCTIONS

Let $A = \{\alpha, \beta, \gamma, \delta\}$, $B = \{a, b, c\}$, and $C = \{1, 2, 3, 4, 5\}$. Consider the function $f : A \to B$ defined by $f(\alpha) = b, f(\beta) = a, f(\gamma) = c$, and $f(\delta) = c$, and the function $g : B \to C$ defined by $g(a) = 3, g(b) = 1$, and $g(c) = 4$. They are visualized in Figure 6.3.

Viewing the diagram in Figure 6.3, we can follow the paths starting in A and ending in C to create a new function h from A to C. Start with $\alpha \in A$. Since f maps α to b and g maps b to 1, the function h will map α to 1. Now consider $\beta \in A$. Since f maps β to a and g maps a to 3,

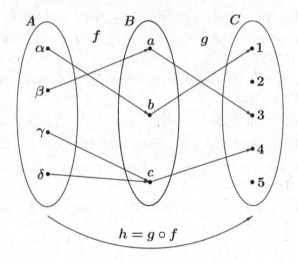

Figure 6.3 Composition of f with g.

the function h will map β to 3. Next look at $\gamma \in A$. Given that f maps γ to c and g maps c to 4, the function h will map γ to 4. Finally, let us track $\delta \in A$. Since f maps δ to c and g maps c to 4, the function h will map δ to 4. In summation, we have $h : \{\alpha, \beta, \gamma, \delta\} \to \{1, 2, 3, 4, 5\}$ defined by $h(\alpha) = 1, h(\beta) = 3, h(\gamma) = 4$, and $h(\delta) = 4$.

Examining the new function h from a different angle, it was determined by plugging each element $x \in A$ into f to obtain $f(x)$, and then plugging $f(x)$ into g to get $g(f(x))$. In other words, for each element in A, we determined h as follows:

$$h(\alpha) = g(f(\alpha)) = g(b) = 1$$
$$h(\beta) = g(f(\beta)) = g(a) = 3$$
$$h(\gamma) = g(f(\gamma)) = g(c) = 4$$
$$h(\delta) = g(f(\delta)) = g(c) = 4$$

As will be defined in the next paragraph, the function h is referred to as the **composition** of the functions f and g and is denoted by $h = g \circ f$.

Definition 6.4.1. *Let A, B, and C be sets, and let $f : A \to B$ and $g : B \to C$ be functions. The **composition** of f with g is the function $g \circ f : A \to C$ defined by $(g \circ f)(a) = g(f(a))$ for all $a \in A$.*

A comment follows before we look at some examples. Writing the composition of functions f with g as $g \circ f$ stems from the fact that when we evaluate a function at an element, the element is written to the right of the function (right-hand notation). But given that we read from left to right in English, the composition $g \circ f$ appears to indicate that we evaluate the function g first. But when we plug an element a into $g \circ f$, we are computing $(g \circ f)(a) = g(f(a))$ so the notation makes sense.

Example 6.4.2. *We will determine the compositions of the following functions.*

(i) *Consider the function $f : \{1,2,3\} \to \{\alpha, \beta, \gamma, \delta\}$ defined by $f = \{(1,\alpha), (2,\gamma), (3,\beta)\}$ and $g : \{\alpha, \beta, \gamma, \delta\} \to \{a,b,c,d\}$ defined by $g = \{(\alpha,d), (\beta,c), (\gamma,a)\}$. Then the composition of f with g is the function $g \circ f : \{1,2,3\} \to \{\alpha, \beta, \gamma, \delta\}$ defined by*

$$(g \circ f)(1) = g(f(1)) = g(\alpha) = d$$
$$(g \circ f)(2) = g(f(2)) = g(\gamma) = a, \text{ and}$$
$$(g \circ f)(3) = g(f(3)) = g(\beta) = c.$$

(ii) *Consider the function $f : \mathbb{R} \to \mathbb{R}$ defined by $f(x) = x - 3$ and the function $g : \mathbb{R} \to \mathbb{R}$ defined by $g(x) = x^2 + 7x + 2$. Then the composition of f with g is the function $g \circ f : \mathbb{R} \to \mathbb{R}$ defined by*

$$
\begin{aligned}
(g \circ f)(x) &= g(f(x)) \\
&= g(x - 3) \\
&= (x - 3)^2 + 7(x - 3) + 2 \\
&= x^2 - 6x + 9 + 7x - 21 + 3 \\
&= x^2 + x - 9.
\end{aligned}
$$

Now in this example, we can also compute $f \circ g$. This composition is defined by

$$
\begin{aligned}
(f \circ g)(x) &= f(g(x)) \\
&= f(x^2 + 7x + 2) \\
&= (x^2 + 7x + 2) - 3 \\
&= x^2 + 7x - 1.
\end{aligned}
$$

Example 6.4.2 brings up two important points. First note that if you can compute $g \circ f$ for two functions f and g, you can't necessarily compute $f \circ g$. In Example 6.4.2 (i), you would not be able to compute $f \circ g$. Now this was possible in Example 6.4.2 (ii) as both functions have the same domain and codomain. However, note that as functions $g \circ f \neq f \circ g$, which tells us that the operation of composition of functions is not commutative. The following results state two properties that composition of functions does satisfy.

Lemma 6.4.3. *Let $f : A \to B$ be a function. Then the identity function i_B satisfies $i_B \circ f = f$ and $f \circ i_A = f$.*

The proof is an exercise.

Lemma 6.4.4. *For sets A, B, C and D with function $f : A \to B, g : B \to C$, and $h : C \to D$, then $(h \circ g) \circ f = h \circ (g \circ f)$.*

In other words, the operation of composition of function is associative.

Proof of Lemma 6.4.4. Let $a \in A$. We must show that $((h \circ g) \circ f)(a) = (h \circ (g \circ f))(a)$. Given that

$$((h \circ g) \circ f)(a) = (h \circ g)(f(a)) = h(g(f(a)))$$

and

$$(h \circ (g \circ f))(a) = h((g \circ f)(a)) = h(g(f(a))),$$

it follows that $(h \circ g) \circ f = h \circ (g \circ f)$. $\qquad\square$

There are other properties that composition of function satisfies.

Theorem 6.4.5. *Let A, B, and C be sets, and let $f : A \to B$ and $g : B \to C$ be functions.*

(i) *If f and g are both one-to-one, then $g \circ f$ is one-to-one.*

(ii) *If f and g are both onto, then $g \circ f$ is onto.*

(iii) *If f and g are both bijections, then $g \circ f$ is a bijection.*

We will only prove (i) from Theorem 6.4.5. The proofs of the other two statements are exercises.

Proof of Theorem 6.4.5 (i). First note that as a function, we have $g \circ f$: $A \to C$. To show that $g \circ f$ is one-to-one, let $a_1, a_2 \in A$ and assume $(g \circ f)(a_1) = (g \circ f)(a_2)$. This implies that $g(f(a_1)) = g(f((a_2))$. Given that g is one-to-one, this means that $f(a_1) = f(a_2)$. But f is also one-to-one, yielding $a_1 = a_2$. Since assume $(g \circ f)(a_1) = (g \circ f)(a_2)$ resulted in $a_1 = a_2$, we have that $g \circ f$ is one-to-one. □

Exercises 6.4

(1) For each pair of functions, compute the indicated composition.

    (a) Let $A = \{0, 2, 4, 6, 8\}, B = \{a, b, c, d\}$, and $C = \{-1, -2, -3, -4, -5\}$. Define functions $f : A \to B$ by $f = \{(0, c), (2, d), (4, d), (6, b), (8, a)\}$ and $g : B \to C$ by $g = \{(a, -1), (b, -3), (c, -4), (d, -1)\}$. Find $g \circ f$.

    (b) Consider the function $f : \mathbb{Z} \to \mathbb{Z}$ defined by $f(n) = 1 - n$ for $n \in \mathbb{Z}$ and the function $g : \mathbb{Z} \to \mathbb{N}$ defined by $g(m) = m^2$ for $m \in \mathbb{Z}$. Find $g \circ f$.

    (c) Consider the function $f : \mathbb{Z} \to \mathbb{Z}$ defined by $f(n) = 2n + 3$ for $n \in \mathbb{Z}$ and the function $g : \mathbb{Z} \to \mathbb{Z}$ defined by $g(m) = 2$ for $m \in \mathbb{Z}$. Find $g \circ f$ and $f \circ g$.

    (d) Consider the function $f : \mathbb{Z} \times \mathbb{Z} \to \mathbb{Z}$ defined by $f(n, m) = n + m$ for $(n, m) \in \mathbb{Z} \times \mathbb{Z}$ and the function $g : \mathbb{Z} \to \mathbb{Z} \times \mathbb{Z}$ defined by $g(t) = (-t, t)$ for $t \in \mathbb{Z}$. Find $g \circ f$ and $f \circ g$.

    (e) Consider the function $f : \mathbb{R} \to \mathbb{R}$ defined by $f(x) = x - 1$ for $x \in \mathbb{R}$ and the function $g : \mathbb{R} \to \mathbb{R}$ defined by $g(x) = x^2 - 1$ for $x \in \mathbb{R}$. Find $g \circ f, f \circ g, g \circ g$, and $f \circ f$.

    (f) Consider the function $f : \mathbb{R}^* \to \mathbb{R}^*$ defined by $f(x) = \frac{2}{x}$ for $x \in \mathbb{R}^*$ and the function $g : \mathbb{R}^* \to \mathbb{R}^*$ defined by $g(x) = \frac{1}{2x}$ for $x \in \mathbb{R}^*$. Find $g \circ f, f \circ g, g \circ g$, and $f \circ f$.

    (g) Consider the function $f : \mathbb{R} \to \mathbb{R}$ defined by $f(x) = 2x$ for $x \in \mathbb{R}$, the function $g : \mathbb{R} \to \mathbb{R}$ defined by $g(x) = x^2$ for $x \in \mathbb{R}$, and the function $h : \mathbb{R} \to \mathbb{R}$ defined by $h(x) = \sqrt{x^2 + 1}$. Find $h \circ g \circ f, f \circ f \circ f, f \circ g \circ h$, and $h \circ h \circ g$.

(2) Prove Theorem 6.4.5 (ii): for functions $f : A \to B$ and $g : B \to C$, prove that if f and g are both onto, then $g \circ f$ is onto.

(3) Prove Theorem 6.4.5 (iii): for functions $f : A \to B$ and $g : B \to C$, prove that if f and g are both bijections, then $g \circ f$ is a bijection.

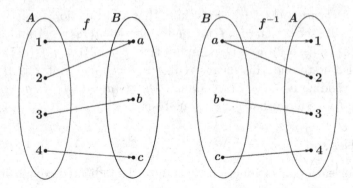

Figure 6.4 Function f and its inverse from Example 6.5.1.

(4) Let $f : A \to B$ and $g : B \to C$ be functions.

   (a) If $g \circ f$ is one-to-one, prove that f is one-to-one.

   (b) If $g \circ f$ is one-to-one and f is onto, prove that g is one-to-one.

   (c) If $g \circ f$ is onto, prove that g is onto.

   (d) If $g \circ f$ is onto and g is one-to-one, prove that f is onto.

(5) Prove Lemma 6.4.3: For a function $f : A \to B$, the identity function i_B satisfies $i_B \circ f = f$ and $f \circ i_A = f$.

6.5   INVERSE OF A FUNCTION

Given a relation R from a set A to a set B, we can define the inverse relation R^{-1} from B to A by $b\,R^{-1}\,a$ if and only if $a\,R\,b$ (see Definition 5.1.3). Given that a function is a relation, we can compute the inverse of a function.

Example 6.5.1. *Consider the sets* $A = \{1, 2, 3, 4\}, B = \{a, b, c\}, C = \{a, b, c, d\}$, *and* $D = \{a, b, c, d, e\}$

   (i) *Define the function* $f : A \to B$ *by* $f = \{(1, a), (2, a), (3, b), (, 4, c)\}$. *The inverse* f^{-1} *from* B *to* A *is the relation* $f^{-1} = \{(a, 1), (a, 2), (b, 3), (c, 4)\}$. *The function* f *and the relation* f^{-1} *are visualized in Figure 6.4.*

   The diagram shows that while f *is a function, its inverse is not as* f^{-1} *relates* a *in* B *to two different elements in* A. *The first*

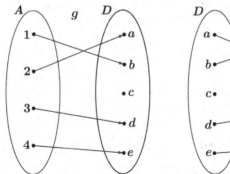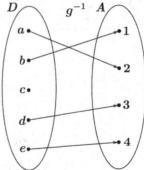

Figure 6.5 Function g and its inverse from Example 6.5.1.

lesson is that the inverse of a function may not be a function. Here, this is due to the fact that f is not one-to-one. Given that $f(1) = f(2) = a$, the inverse relation f^{-1} relates a to both 1 and 2.

(ii) *Define the function $g : A \to D$ by $g = \{(1, b), (2, a), (3, d), (4, e)\}$. The inverse g^{-1} from D to A is the relation $g^{-1} = \{(b, 1), (a, 2), (d, 3), (e, 4)\}$. The function g and the relation g^{-1} are visualized in Figure 6.5.*

The diagram shows that while g is a function, its inverse g^{-1} is not as g^{-1} does not relate the element c in D to any element in A. This is due to the fact that g is not onto. Given that $c \notin ran(g)$, the inverse relation g^{-1} does not relate c to any element in A.

(iii) *Define the function $h : A \to C$ by $h = \{(1, a), (2, d), (3, b), (4, c)\}$. The inverse $h^{-1} : C \to A$ is the relation $h^{-1} = \{(a, 1), (b, 3), (c, 4), (d, 2)\}$. The function h and the relation h^{-1} are visualized in Figure 6.6. The function h is one-to-one and onto (a bijection), and note that h^{-1} is also a function.*

The functions from Example 6.5.1 motivate the following result.

Theorem 6.5.2. *Let $f : A \to B$ be a function. The inverse relation f^{-1} is a function if and only if f is a bijection.*

Proof. ($\Rightarrow$) Assume that f^{-1} is a function. First we will show that f is one-to-one. Let $a_1, a_2 \in A$ and assume that $f(a_1) = f(a_2)$. Note that

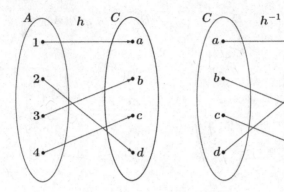

Figure 6.6 Inverse of function h.

$f(a_1)$ and $f(a_2)$ are elements in B with $f(a_1) = f(a_2) = b \in B$. Given that $f^{-1} : B \to A$ is a function, f^{-1} maps every element in B to an element if A. This means that $f^{-1}(b)$ is defined. Furthermore, by the definition of an inverse relation, $f^{-1}(b) = a$ where $a \in A$ and $f(a) = b$. Given that $f(a_1) = b$ and $f(a_2) = b$, we have that $f^{-1}(b) = a_1$ and $f^{-1}(b) = a_2$. But again, given that f^{-1} is a function, it can't map a single element in B to two distinct elements in A. Therefore, $a_1 = a_2$ and f is one-to-one.

To show that f is onto, let $b \in B$. Since f^{-1} is a function from B to A, $f^{-1}(b) = a$ where a is the unique element in A such that $f(a) = b$. This implies that f is onto. Given that f is one-to-one and onto, it is a bijection.

($\Leftarrow$) For the reverse direction, assume that f is a bijection. To show that f^{-1} is a function from B to A, we must show that f^{-1} relates every element in B to exactly one element in A. Let $b \in B$. Since $f : A \to B$ is onto, there exists an element $a \in A$ such that $f(a) = b$. By the definition of the inverse of a relation, $f^{-1}(b) = a$. Suppose there exists another element $c \in A$ such that $f^{-1}(b) = c$. This would imply that $f(a) = b$ and that $f(c) = b$, implying that $f(a) = f(c)$. But since f is one-to-one, we get that $a = c$. Since f^{-1} relates every element in B to exactly one element in A, f^{-1} is a function. $\qquad\square$

For the moment, let's return to function h from Example 6.5.1. Recall that $h : A \to C$, where $A = \{1, 2, 3, 4\}$ and $C = \{a, b, c, d\}$, was defined by $h(1) = a, h(2) = d, h(3) = b$, and $h(4) = c$. The inverse of h was the function $h^{-1} : C \to A$ defined by $h^{-1}(a) = 1, h^{-1}(b) = 3, h^{-1}(c) =$

4, and $h^{-1}(d) = 2$. Computing the function $h^{-1} \circ h : \{1, 2, 3, 4\} \to \{1, 2, 3, 4\}$, we get the following:

$$(h^{-1} \circ h)(1) = h^{-1}(h(1)) = h^{-1}(a) = 1$$
$$(h^{-1} \circ h)(2) = h^{-1}(h(2)) = h^{-1}(d) = 2$$
$$(h^{-1} \circ h)(3) = h^{-1}(h(3)) = h^{-1}(b) = 3$$
$$(h^{-1} \circ h)(4) = h^{-1}(h(4)) = h^{-1}(c) = 4$$

We notice the composition of h^{-1} with h sends every element in $A = \{1, 2, 3, 4\}$ to itself or that $h^{-1} \circ h = i_A$. Next we will examine the composition $h \circ h^{-1} : \{a, b, c, d\} \to \{a, b, c, d\}$. A quick calculation shows the following:

$$(h \circ h^{-1})(a) = h(h^{-1}(a)) = h(1) = a$$
$$(h \circ h^{-1})(b) = h(h^{-1}(b)) = h(3) = b$$
$$(h \circ h^{-1})(c) = h(h^{-1}(c)) = h(4) = c$$
$$(h \circ h^{-1})(d) = h(h^{-1}(d)) = h(2) = d$$

Here, the composition sends every element in the set $C = \{a, b, c, d\}$ to itself or $h \circ h^{-1} = i_C$. Now the formal definition of the inverse of a function can be stated.

Definition 6.5.3. *Let the function $f : A \to B$ be a bijection. The* **inverse** *of f is the function $f^{-1} : B \to A$ such that $f^{-1} \circ f = i_A$ and $f \circ f^{-1} = i_B$.*

Example 6.5.4.

(i) *Consider the function $f : \mathbb{Q} \to \mathbb{Q}$ defined by $f(r) = 2r + 1$ and the function $g : \mathbb{Q} \to \mathbb{Q}$ defined by $g(r) = \frac{r-1}{2}$. First, we compute $g \circ f$ for $r \in \mathbb{Q}$.*

$$(g \circ f)(r) = g(f(r)) = g(2r + 1) = \frac{2r + 1 - 1}{2} = \frac{2r}{2} = r = i_{\mathbb{Q}}(r)$$

This implies that $g \circ f = i_{\mathbb{Q}}$. Next, we compute $f \circ g$ for $r \in \mathbb{Q}$.

$$(f \circ g)(r) = f(g(r)) = f\left(\frac{r-1}{2}\right) = 2 \cdot \frac{r-1}{2} + 1$$

$$= r - 1 + 1 = r = i_{\mathbb{Q}}(r)$$

This implies that $f \circ g = i_{\mathbb{Q}}$. Therefore, the functions f and g are inverse to each other.

(ii) *This simple example shows that a function can be its own inverse. Consider the function $f : \mathbb{Z} \to \mathbb{Z}$ defined by $f(n) = -n$. Now compute $f \circ f$ for $n \in \mathbb{Z}$.*

$$(f \circ f)(n) = f(f(n)) = f(-n) = -(-n) = n = i_{\mathbb{Z}}(n)$$

This implies that $f \circ f = i_{\mathbb{Z}}$. Therefore, f is its own inverse.

Recall that in Theorem 6.5.2 we proved that the inverse of function is also a function if and only if the original function is a bijection. At this point we prove a stronger version of that result.

Theorem 6.5.5. *Every bijective function has a unique inverse function that is also a bijection.*

Proof. Let $f : A \to B$ be a bijective function. We know by Theorem 6.5.2, that f has an inverse function. Suppose we have two functions $g : B \to A$ and $h : B \to A$ that are inverses of f. This means that $g \circ f = i_A$ and that $f \circ h = i_B$.

First note by Lemma 6.4.3 that $g = g \circ i_B$. Given that $f \circ h = i_B$, we now have

$$g = g \circ i_B = g \circ (f \circ h).$$

By Lemma 6.4.4, we know that the operation of composition of functions is associative. This means

$$g = g \circ (f \circ h) = (g \circ f) \circ h.$$

Apply the fact that $g \circ f = i_A$ results in

$$g = (g \circ f) \circ h = i_A \circ h.$$

Using Lemma 6.4.3 once more, we arrive at

$$g = i_A \circ h = h.$$

Thus $g = h$ and the inverse of f is unique.

To show that the inverse of a bijective function $f : A \to B$ is also a bijection, let $f^{-1} : B \to A$ be the inverse of f. To show it is one-to-one, let $b_1, b_2 \in B$ and assume that $f^{-1}(b_1) = f^{-1}(b_2)$. This implies that $f(f^{-1}(b_1)) = f(f^{-1}(b_2))$ or that $(f \circ f^{-1})(b_1) = (f \circ f^{-1})(b_2)$. Given that $f \circ f^{-1} = i_B$, it follows that $(f \circ f^{-1})(b_1) = (f \circ f^{-1})(b_2)$ implies $i_B(b_1) = i_B(b_2)$. By the definition of the identity function, we have that $b_1 = b_2$ and f^{-1} is one-to-one.

To show f^{-1} is onto, let $a \in A$. We know that $f(a) = b$ for some $b \in B$. This implies that $f^{-1}(b) = f^{-1}(f(a)) = (f^{-1} \circ f)(a) = i_A(a)$. By the definition of the identity function, we have that $f^{-1}(b) = a$ and f^{-1} is onto. Since f^{-1} is one-to-one and onto, it is a bijection. $\qquad\square$

So how do we determine the inverse of a function $f : A \to B$ when we know one exists? Given that $f^{-1} : B \to A$ exists, we know by Definition 6.5.3 that $f \circ f^{-1} = i_B$. This means for all $b \in B$ that $(f \circ f^{-1})(b) = i_B(b)$ or that $f(f^{-1}(b)) = b$. Then, if possible, solve for $f^{-1}(b)$. We end this chapter by using this method to find the inverse of the function $f : \mathbb{R} - \{2\} \to \mathbb{R} - \{3\}$ defined by $f(x) = \frac{3x}{x-2}$ from Example 6.3.3.

Let $x \in \mathbb{R} - \{3\}$. We know that $(f \circ f^{-1})(x) = i_{\mathbb{R}-\{3\}}(x)$ or that $f(f^{-1}(x)) = x$. Now solve for $f^{-1}(x)$.

$$f(f^{-1}(x)) = x$$
$$\frac{3f^{-1}(x)}{f^{-1}(x) - 2} = x$$
$$3f^{-1}(x) = x(f^{-1}(x) - 2)$$
$$3f^{-1}(x) = xf^{-1}(x) - 2x$$
$$3f^{-1}(x) - xf^{-1}(x) = -2x$$
$$f^{-1}(x)(3 - x) = -2x$$
$$f^{-1}(x) = \frac{-2x}{3 - x} = \frac{2x}{x - 3}$$

Exercises 6.5

(1) Determine which of the following functions are invertible.

   (a) Let $A = \{0, 2, 4, 6, 8\}$ and $B = \{a, b, c, e, d\}$

      i. Function $f : A \to B$ defined by $f = \{(0, e), (2, d), (4, c), (6, b), (8, a)\}$.

      ii. Function $g : A \to B$ defined by $g = \{(0, d), (2, c), (4, b), (6, a), (8, d)\}$.

   (b) Function $f : \mathbb{Z} \times \mathbb{Z} \to \mathbb{Z}$ defined by $f(n, m) = n + m + 3$ for $(n, m) \in \mathbb{Z} \times \mathbb{Z}$.

(c) Function $f : \mathbb{Z} \times \mathbb{Z} \to \mathbb{Z}$ defined by $f(n, m) = (-1)^n m$ for $(n, m) \in \mathbb{Z} \times \mathbb{Z}$.

(d) Function $f : \mathbb{R} \to \mathbb{R}$ defined by $f(x) = x^2 - 4$ for $x \in \mathbb{R}$.

(e) Function $f : \mathbb{R} - \{-7\} \to \mathbb{R} - \{5\}$ defined by $f(x) = \frac{5x+1}{x+7}$ for $x \in \mathbb{R}$.

(f) Function $f : \mathbb{R} \to \mathbb{R}$ defined by $f(x) = x$ for $x \in \mathbb{R}$.

(g) Function $f : \mathbb{Z} \times \mathbb{Z} \to \mathbb{Z} \times \mathbb{Z}$ defined by $f(n, m) = (n+m, n-m)$ for $(n, m) \in \mathbb{Z} \times \mathbb{Z}$.

(2) Determine if the following pairs of functions are inverses to each other.

(a) Function $f : \mathbb{Z} \to \mathbb{Z}$ defined by $f(n) = n - 2$ for $n \in \mathbb{Z}$ and function $g : \mathbb{Z} \to \mathbb{Z}$ defined by $g(n) = 2 + n$ for $n \in \mathbb{Z}$.

(b) Function $f : \mathbb{Q} \to \mathbb{Q}$ defined by $f(r) = 5r - 3$ for $r \in \mathbb{Q}$ and function $g : \mathbb{Q} \to \mathbb{Q}$ defined by $g(r) = \frac{r+3}{5}$ for $r \in \mathbb{Q}$.

(c) Function $f : \mathbb{R} \to \mathbb{R}$ defined by $f(x) = x^3 + 1$ for $x \in \mathbb{R}$ and function $g : \mathbb{R} \to \mathbb{R}$ defined by $g(x) = (x - 1)^3$ for $x \in \mathbb{R}$.

(d) Function $f : \mathbb{R} - \{8\} \to \mathbb{R} - \{3\}$ defined by $f(x) = \frac{3x}{x-8}$ for $x \in \mathbb{R}$ and function $g : \mathbb{R} - \{3\} \to \mathbb{R} - \{8\}$ defined by $g(x) = \frac{8x}{x-3}$ for $x \in \mathbb{R}$.

(e) Function $f : \mathbb{R} \to \mathbb{R}^+$ defined by $f(x) = 3e^{x^2}$ for $x \in \mathbb{R}$ and function $g : \mathbb{R}^+ \to \mathbb{R}$ defined by $g(x) = \ln\left(\frac{\sqrt{x}}{3}\right)$ for $x \in \mathbb{R}$.

(f) Function $f : \mathbb{R} \to \mathbb{R}^+ \cup \{0\}$ defined by $f(x) = x^2$ for $x \in \mathbb{R}$ and function $g : \mathbb{R}^+ \cup \{0\} \to \mathbb{R}$ defined by $g(x) = \sqrt{x}$ for $x \in \mathbb{R}$.

(3) Let A be a nonempty set. Prove that the identity function i_A on A is invertible and find its inverse.

(4) Let $f : A \to B$ be an invertible function with inverse f^{-1}. Prove that the following statements are true.

(a) $f^{-1} \circ f \circ f^{-1} = f^{-1}$.

(b) $f \circ f^{-1} \circ f = f$.

(c) $f^{-1} \circ f \circ f^{-1} \circ f = i_A$.

CHAPTER 7

Cardinality of Sets

7.1 INTRODUCTION

The concept of the cardinality of a set refers to the size or the number of elements in that set. At the moment, as was briefly mentioned in Chapter 3, our only grasp of cardinality is the notion of a set being finite or infinite. We know that the set $\{a, b, c, d\}$ is finite. There are four elements in this set or the size of the set is 4. If we consider the set $\mathbb{Q}$, we would say that $\mathbb{Q}$ is infinite, as it does not appear possible for a natural number $n \in \mathbb{N}$ to exist such that number of elements in $\mathbb{Q}$ is n.

The concept of counting the number of elements in a set dates back thousands of years. Archaeological evidence suggests that cultures dating as far back as 8000 BC used small tokens to represent quantities of manufactured products and agricultural goods. Thus by counting tokens, you would be counting the number of objects each token represented. For example, a farmer might use one disc-shaped token to represent each sheep in their flock. Bringing a collection of 12 of these tokens to market would show that they had 12 sheep to sell or trade. The farmer may also grow wheat. So as not to confuse sheep with wheat, a cone-shaped token may have been used to represent one bushel of wheat. Thus a collection of 20 cone-shaped tokens would indicate that the farmer had 20 bushels of wheat to sell or trade. Thus the tokens are used as a way to count the number of elements in a certain set the farmer is interested in.

It has also been suggested that counting (mathematics) is the precursor to writing. Evidence has been found of clay tablets with a series of token impressions on them. Perhaps this was done to have a single object, the clay tablet, with a list of the agricultural products the farmer has available for sale or trade. It has been suggested that the method of pushing tokens into wet clay to create an imprint led to the cuneiform system of writing. If you are interested in this subject, I suggest the book [SB92].

7.2   SETS WITH THE SAME CARDINALITY

As we mention in Section 1 of this chapter, the set $A = \{a, b, c, d\}$ has four elements. However, the set $B = \{1, 2, 3, 4\}$ also has four elements, while the set $C = \{\alpha, \beta, \gamma\}$ has three elements and the set $D = \{\square, \triangle, \heartsuit, \diamondsuit, \perp\}$ has five elements. The sets A and B have the same size, while set A does not have the same size as sets C and D.

For sets A and B, we can construct a bijection f from A to B defined by $f = \{(a, 2), (b, 1), (c, 4), (d, 3)\}$. This is visualized in Figure 7.1. Note that f is not unique as a bijective function from A to B. The function $g : A \to B$ defined by $g = \{(a, 1), (b, 3), (c, 2), (d, 4)\}$ is also a bijection visualized in Figure 7.1. The important thing to note here is that there exists a bijection from A to B.

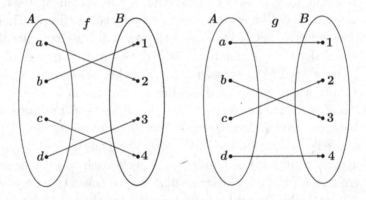

Figure 7.1   Bijections f and g from $\{a, b, c, d\}$ to $\{1, 2, 3, 4\}$.

Now let's look at the sets A and C. There is no way to define a bijection from A to C as any mapping from A to C will never be one-to-one (always contain a collision). For example, consider the function $h : A \to C$ defined by $h(a) = \alpha, h(b) = \beta, h(c) = \gamma$, and $h(d) = \gamma$, which is visualized in Figure 7.2.

Finally, let's look at the sets A and D. There is no way to define a bijection from A to D as the function will never be onto. For example, consider the function $k : A \to D$ defined by $k(a) = \square, k(b) = \triangle, k(c) = \heartsuit$, and $k(d) = \diamondsuit$, which is visualized in Figure 7.2.

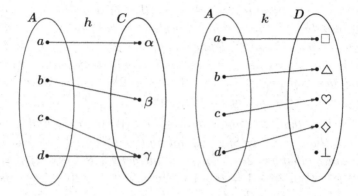

Figure 7.2 Functions h and k.

These examples motivate the following definition.

Definition 7.2.1. *Two sets A and B have the **same cardinality** if and only if there exists a bijection from A to B.*

Note that in the definition, the sets can be finite or infinite. This means that two infinite sets can have the same cardinality.

Example 7.2.2. *We will show that the following pairs of sets have the same cardinality.*

(i) *Let $A = \{-6, -5, -4, -3, -2, -1, 0\}$ and $B = \{1, 2, 3, 4, 5, 6, 7\}$. To show that A and B have the same cardinality, we must find a bijection from A to B. Consider the mapping $f : A \to B$ defined by $f(a) = -a + 1$ for $a \in A$. To show f is one-to-one, let $a_1, a_2 \in A$ and assume that $f(a_1) = f(a_2)$. This implies that $-a_1 + 1 = -a_2 + 1$ or that $-a_1 = -a_2$. Consequently we get that $a_1 = a_2$ and f is one-to-one.*

To show that f is onto, let $b \in B$. We need to find an $a \in A$ such that $f(a) = b$. If $f(a) = b$, then $-a + 1 = b$ or $-a = b - 1$. This yields $a = -b + 1$. A simple calculation shows that $-6 \leq -b + 1 \leq 0$ and that $-b + 1 \in A$. Lastly, given that $f(-b + 1) = -(-b + 1) + 1 = b - 1 + 1 = b$, we have that f is onto. Thus f is a bijection, and A and B have the same cardinality.

(ii) *Consider the sets $\mathbb{N}$ and the set $O^+ = \{1, 3, 5, 7, 9, \ldots\}$ the set* *of positive odd integers. To show that $\mathbb{N}$ and O^+ have the same cardinality, we must find a bijection from $\mathbb{N}$ to O^+. Consider the mapping $g : \mathbb{N} \to O^+$ defined by $g(n) = 2n - 1$ for $n \in \mathbb{N}$. To show g is one-to-one, let $n, m \in \mathbb{N}$ and assume that $g(n) = g(m)$. This implies that $2n - 1 = 2m - 1$ or that $2n = 2m$. Consequently we get that $n = m$ and g is one-to-one.*

To show that g is onto, let $b \in O^+$. We need to find an $n \in \mathbb{N}$ such that $g(n) = b$. Start by noting that b is a positive odd integer, meaning that $b = 2l+1$ for some integer l with $l \geq 0$. If $g(n) = b = 2l + 1$, then $2n - 1 = 2l + 1$ or $2n = 2l + 2 = 2(l + 1)$. This yields $n = l + 1$. Given that $l \geq 0$, we have $l + 1 \geq 1$ and that $l + 1 \in \mathbb{N}$. Lastly, given that $g(l + 1) = 2(l + 1) - 1 = 2l + 2 - 1 = 2l + 1 = b$, we have that g is onto. Thus g is a bijection and $\mathbb{N}$ and O^+ have the same cardinality.

(iii) *We end with the classic example that the interval $(0, 1)$ and $\mathbb{R}^+ = (0, \infty)$ have the same cardinality. To set up the bijection, draw the interval $(0, 1)$ on the x-axis of the Cartesian plane, and let $x \in (0, 1)$. This means $0 < x < 1$, which implies that $1 < \frac{1}{x}$. The graph of $\frac{1}{x}$, for $0 < x < 1$, is the dashed curve in Figure 7.3. Given that we want a function from $(0, 1)$ to $(0, \infty)$, subtract 1 from this function to obtain the function $\frac{1}{x} - 1 = \frac{1-x}{x}$. The graph of this function, for $0 < x < 1$, is the solid curve in Figure 7.3.*

Putting all of this together, define a function $f : (0, 1) \to (0, \infty)$ by $f(x) = \frac{1-x}{x}$ for $x \in (0, 1)$. To show f is one-to-one, let $x_1, x_2 \in (0, 1)$ and assume that $f(x_1) = f(x_2)$. This implies that $\frac{1-x_1}{x_1} = \frac{1-x_2}{x_2}$. The calculation

$$\frac{1 - x_1}{x_1} = \frac{1 - x_2}{x_2}$$
$$x_1 - x_1 x_2 = x_2 - x_1 x_2$$
$$x_1 = x_2$$

shows that f is one-to-one.

To show f is onto, let $y \in (0, \infty)$ and consider the element $x = \frac{1}{y+1}$. Given that $0 < y$, it follows that $1 < y+1$ and that $0 < \frac{1}{y+1} < 1$. Evaluating, we obtain

$$f\left(\frac{1}{y+1}\right) = \frac{1 - \frac{1}{y+1}}{\frac{1}{y+1}}$$

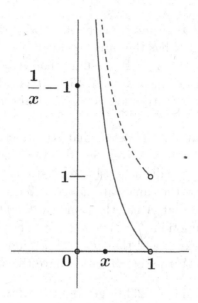

Figure 7.3 Bijection from $(0, 1)$ to $(0, \infty)$.

$$= y + 1 - 1$$
$$= y.$$

Thus f is onto. Now that we have shown f is a bijection, we have that the interval $(0, 1)$ and $\mathbb{R}^+ = (0, \infty)$ have the same cardinality.

Before moving on, we state a result concerning some of the basic properties of cardinality.

Lemma 7.2.3. *Let $A, B,$ and C be sets.*

(i) *A has the same cardinality as $\emptyset$ if and only if $A = \emptyset$.*

(ii) *A has the same cardinality as A.*

(iii) *If A has the same cardinality as B, then B has the same cardinality as A.*

(iv) *If A has the same cardinality as B and B has the same cardinality as C, then A has the same cardinality as C.*

Only (i) and (iii) are proved here; the others are exercises.

Proof of Lemma 7.2.3 (i) and (iii). First we will prove (i).

($\Rightarrow$) Assume that A has the same cardinality as $\emptyset$. Then there exists a bijective function $f : A \to \emptyset$. Assume that A is not the empty set. Then there exists an element $a \in A$. Since f is a function, $f(a)$ is an element in $\emptyset$. This is a contraction. Thus there are no elements in A and $A = \emptyset$.

($\Leftarrow$) Assume that $A = \emptyset$. Recall that any function from A to $\emptyset$ will be a subset of $A \times \emptyset = \emptyset \times \emptyset$. Given that $\emptyset \times \emptyset = \emptyset$ has only $\emptyset$ as a subset, let f be that subset. It will vacuously be one-to-one and onto. Thus f is a bijection and A has the same cardinality as $\emptyset$.

For (iii), assume that A has the same cardinality as B. Thus there exists a bijective function $f : A \to B$. By Theorem 6.5.5, we know that $f^{-1} : B \to A$ exists and that it is a bijection. Since there exists a bijection from B to A, the sets B and A have the same cardinality. $\square$

We can apply Lemma 7.2.3 to get the following result.

Theorem 7.2.4. *The interval $(0,1)$ has the same cardinality as the real numbers $\mathbb{R}$.*

Proof. We know from Example 7.2.2 (iii) that the interval $(0,1)$ and $\mathbb{R}^+$ have the same cardinality. As an exercise, you will show that $\mathbb{R}$ and $\mathbb{R}^+$ have the same cardinality. This means by Lemma 7.2.3 that $\mathbb{R}^+$ and $\mathbb{R}$ have the same cardinality. Since the interval $(0,1)$ and $\mathbb{R}^+$ have the same cardinality and $\mathbb{R}^+$ and $\mathbb{R}$ have the same cardinality, it follows from Lemma 7.2.3 that the interval $(0,1)$ and $\mathbb{R}$ have the same cardinality. $\square$

Exercises 7.2

(1) Show that the following pairs of sets have the same cardinality. Be sure to justify your answer.

(a) $\{1, 2, \ldots, n\}$ and $\{-1, -2, \ldots, -n\}$ for $n \in \mathbb{N}$.

(b) $\mathbb{R}^+$ and $\mathbb{R}^-$.

(c) $\mathbb{Z}$ and E, where E is the set of even integers.

(d) $\mathbb{Z}$ and O, where O is the set of odd integers.

(e) $\mathbb{R}$ and $\mathbb{R}^+$.

(f) $\mathbb{R}$ and the interval $(5, \infty)$.

(g) $\mathbb{Z}$ and $\{\ldots, \frac{1}{27}, \frac{1}{9}, \frac{1}{3}, 1, 3, 9, 27, \ldots\}$.

(h) The interval $(0,1)$ and the interval (a, b) with $a, b \in \mathbb{R}$.

    (i) $\{1, 2\} \times \mathbb{N}$ and $\mathbb{Z}$.

(2) Let A and B be sets with the same cardinality, prove that $\mathcal{P}(A)$ has the same cardinality as $\mathcal{P}(B)$.

(3) Given intervals (a, b) and (c, d), with $a, b, c, d \in \mathbb{R}$, prove that (a, b) and (c, d) have the same cardinality.

(4) Prove Lemma 7.2.3 (ii): for A a set, A has the same cardinality as A.

(5) Prove Lemma 7.2.3 (iv): for sets A, B and C, if A has the same cardinality as B and B has the same cardinality as C, then A has the same cardinality as C.

(6) For any element x and set A, prove that $\{x\} \times A$ and A have the same cardinality.

7.3 FINITE AND INFINITE SETS

We now proceed to formally defining the concept of finite and infinite sets.

Definition 7.3.1. *Let A be a set.*

    (i) *The set A is **finite** if and only if it is the empty set or it has the same cardinality as the set $\{1, 2, \cdots, n\}$ for some $n \in \mathbb{N}$.*

    (ii) *The set A is **infinite** if and only if it is not finite.*

When A is finite, we can compute the **order** or **cardinality** of A, denoted by $|A|$. If $A = \emptyset$, then $|A| = 0$. When A has the same cardinality as the set $\{1, 2, \ldots, n\}$, for $n \in \mathbb{N}$, then $|A| = n$. This notation is borrowed for infinite sets as well to denote when two sets have the same cardinality. For example, the conclusion from Theorem 7.2.4, that the interval $(0, 1)$ has the same cardinality as the real numbers $\mathbb{R}$, can be written as $|(0, 1)| = |\mathbb{R}|$.

Now let's stick with finite sets for a moment. Suppose that a finite set A has order n for $n \in \mathbb{N}$. Thus there exists a bijection $f : \{1, 2, \ldots, n\} \to A$. Since f is onto, for every element $a \in A$, there exists an $i \in \{1, 2, \ldots, n\}$ such that $f(i) = a$. By denoting $a = f(i) = a_i$, this implies that every element of A can be written in the form a_j, for $j \in \{1, 2, \ldots, n\}$. Furthermore, the representation is

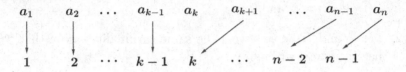

Figure 7.4 Function $g : A - \{a_k\} \to \{1, 2, \ldots, n - 1\}$.

unique. If $a_i = a_j$, then we have $f(i) = f(j)$. But given that f is one-to-one, this implies $i = j$. This means we can write A as $\{a_1, a_2, \ldots, a_n\}$ or $A = \{a_i : i \in \{1, 2, \ldots, n\}\}$ where $f(i) = a_i$. This will come in handy when proving results concerning finite sets.

Lemma 7.3.2. *Let A be a set such that $|A| = n$.*

(i) *For any element $c \in A$, $|A - \{c\}| = n - 1$.*

(ii) *For any element $d \notin A$, $|A \cup \{d\}| = n + 1$.*

We will only prove (i) from Lemma 7.3.2 as the proof of (ii) is an exercise.

Proof of Lemma 7.3.2 (i). Given that $c \in A$, we have that $A \neq \emptyset$. Suppose $|A| = 1$. This implies $A = \{c\}$ with $A - \{c\} = \emptyset$. Thus $|A - \{c\}| = 0$ and the result holds.

Proceed assuming $|A| = n$, for $n \in \mathbb{N}$ with $n \geq 2$. This means there exists a bijective function $f : A \to \{1, 2, \ldots, n\}$. Given that $A = \{a_1, a_2, \ldots, a_n\}$ with $c \in A$, there exists an integer k, for $1 \leq k \leq n$, such that $c = a_k$. In other words, $A = \{a_1, a_2, \ldots, a_k, \ldots, a_n\}$ with $a_k = c$. Define a function $g : A - \{c\} \to \{1, 2, \ldots, n - 1\}$ by

$$g(a) = \begin{cases} f(a) & \text{if } 1 \leq f(a) \leq k - 1 \\ f(a) - 1 & \text{if } k + 1 \leq f(a) \leq n \end{cases}$$

To help understand the function g being defined, see the visualization in Figure 7.4.

To show that g is one-to-one, let $a, b \in A - \{c\}$ and assume that $g(a) = g(b)$. There are three cases to examine:

Case (1) $1 \leq f(a), f(b) \leq k - 1$. In this case, $g(a) = g(b)$ implies that $f(a) = f(b)$. Since f is one-to-one, we get that $a = b$.

Case (2) $k + 1 \leq f(a), f(b) \leq n$. In this case, $g(a) = g(b)$ implies that $f(a) - 1 = f(b) - 1$. Thus we again have $f(a) = f(b)$. Since f is one-to-one, we get that $a = b$.

Case (3) Without loss of generality, assume that $1 \leq f(a) \leq k-1$ and $k + 1 \leq f(b) \leq n$. In this case, $g(a) = g(b)$ implies that $f(a) = f(b) - 1$. But since $k + 1 \leq f(b) \leq n$, it follows that $k \leq f(b) - 1 \leq n - 1$. This is a contradiction as $1 \leq f(a) \leq k - 1$. Thus this case can't occur.

To show that g is onto, let $t \in \{1, 2, \ldots, n - 1\}$. There are two cases to examine.

Case (1) $1 \leq t \leq k-1$. In this case, consider the element $a_t \in A - \{c\}$. It then follows by the definition of g that $g(a_t) = f(a_t) = t$.

Case (2) $k \leq t \leq n - 1$. This implies that $k + 1 \leq t + 1 \leq n$. In this case, consider the element $a_{t+1} \in A - \{c\}$. It then follows by the definition of g that $g(a_{t+1}) = f(a_{t+1}) - 1 = (t + 1) - 1 = t$.

This proves that h is a bijection and that $|A - \{c\}| = n - 1$. □

The next result follows directly from Lemma 7.3.2 and its proof is an exercise.

Theorem 7.3.3. *For sets A and B, with $A \subseteq B$ and B finite, then A is finite and $|A| \leq |B|$.*

Lemma 7.3.4. *For finite sets A and B, $|A \times B| = |A| \cdot |B|$.*

Proof. If A or B is the empty set, then $A \times B$ is also the empty set. Assume without loss of generality that $A = \emptyset$. Then $|A| = 0$ and $|A| \cdot |B| = 0 \cdot |B| = 0$. Given that $A \times B$ is the empty set, we have that $|A \times B| = 0$ and the result follows.

Assume now that both A and B are nonempty with $|A| = n$ and $|B| = m$, for $n, m \in \mathbb{N}$. We can write A as $\{a_1, a_2, \ldots, a_n\}$ and B as $\{b_1, b_2, \ldots, b_m\}$. Now let $(a_i, b_j) \in A \times B$ for $1 \leq i \leq n$ and $1 \leq j \leq m$. Define a mapping $h : A \times B \to \{1, 2, \ldots, nm\}$ by $h(a_i, b_j) = (i-1)m+j$. The motivation for this function comes from arranging the elements in $A \times B$ in a matrix, as shown in Figure 7.5, and reading across the rows.

To show that h is one-to-one, let $(a_i, b_j), (a_l, b_k) \in A \times B$, with $1 \leq i, l \leq n$, and $1 \leq j, k \leq m$, and assume that $h(a_i, b_j) = h(a_l, b_k)$. This implies $(i-1)m+j = (l-1)m+k$. The following calculation shows

$$(i - 1)m + j = (l - 1)m + k$$
$$(i - 1)m - (l - 1)m = k - j$$
$$(i - l)m = k - j.$$

Given that $1 \leq j \leq m$, we have that $-m \leq -j \leq -1$. The fact that

	b_1	b_2	b_3	$\cdots$	b_m
a_1	(a_1, b_1) 1	(a_1, b_2) 2	(a_1, b_3) 3	$\cdots$	(a_1, b_m) m
a_2	(a_2, b_1) m+1	(a_2, b_2) m+2	(a_2, b_3) m+3	$\cdots$	(a_2, b_m) 2m
a_3	(a_3, b_1) 2m+1	(a_3, b_2) 2m+2	(a_3, b_3) 2m+3	$\cdots$	(a_3, b_m) 3m
$\vdots$	$\vdots$	$\vdots$	$\vdots$		$\vdots$
a_n	(a_n, b_1) (n−1)m+1	(a_n, b_2) (n−1)m+2	(a_n, b_3) (n−1)m+3	$\cdots$	(a_n, b_m) nm

Figure 7.5 Matrix of $A \times B$.

$1 \le k \le m$ implies that $-m + 1 \le k - j \le m - 1$. This means that the only multiple of m that will equal $k - j$ is $0 \cdot m$. Thus the equation $(i - l)m = k - j$ forces $i - l = 0$ or $i = l$. This results in $k - j = 0$ or that $k = j$. This means $(a_i, b_j) = (a_l, b_k)$ and h is one-to-one.

To show that h is onto, let $t \in \{1, 2, \ldots, nm\}$. This implies $0 \le t - 1 \le nm - 1$. By the Division Algorithm (Theorem 2.3.8), we know there exists unique integers q and r such that $t - 1 = qm + r$ where $q \in \mathbb{Z}$ and $0 \le r \le m - 1$. Given that $t - 1$ and m are non-negative, we have that $q \ge 0$. We also know that $q \le n - 1$ or else $qm + r > nm - 1$. This implies $0 \le q \le n - 1$ and that $1 \le q + 1 \le n$. In addition, given that $0 \le r \le m - 1$, we know that $1 \le r + 1 \le m$. Consider the element $a_{q+1} \in A$ and the element $b_{r+1} \in B$. Then $(a_{q+1}, b_{r+1}) \in A \times B$ and

$$
\begin{aligned}
h(a_{q+1}, b_{r+1}) &= ((q + 1) - 1)m + r + 1 \\
&= qm + r + 1 \\
&= t + 1 - 1 \\
&= t.
\end{aligned}
$$

2^k Subsets of $A - \{b\}$	$X \cup \{b\}$ for $X \subseteq A - \{b\}$
$\emptyset$	$\{b\}$
$\{a_1\}$	$\{a_1, b\}$
$\{a_2\}$	$\{a_2, b\}$
$\vdots$	$\vdots$
$\{a_1, a_2, \ldots, a_k\}$	$\{a_1, a_2, \ldots, a_k, b\}$

Figure 7.6 Subsets of A.

This shows that h is onto. Thus h is a bijection and $|A \times B| = |A| \cdot |B| = nm$. □

Now on to another important result motivated by an observation we made in Section 3.2.

Theorem 7.3.5. *For A a finite set, $|\mathcal{P}(A)| = 2^{|A|}$.*

Proof. We proceed by induction on the cardinality of A. Since a finite set has order 0 or order n for $n \in \mathbb{N}$, let $P(n)$ be the statement that $|\mathcal{P}(A)| = 2^{|A|}$ for $n \in \mathbb{Z}$ with $n \geq 0$.

Base Case: We must show that $P(0)$ is true. When $n = 0$, we have that $|A| = 0$, which means by Lemma 7.2.3 that $A = \emptyset$. We know that $\mathcal{P}(\emptyset) = \{\emptyset\}$, implying that $|\mathcal{P}(\emptyset)| = 1$. Given that $2^0 = 1$, $P(0)$ is true.

Inductive Case: Let $k \in \mathbb{Z}$, such that $k \geq 0$, and assume $P(k)$ is true. Thus for any set B with $|B| = k$, we have $|\mathcal{P}(B)| = 2^{|B|} = 2^k$.

To show $P(k+1)$ is true, let A be a set such that $|A| = k+1$. Since $k \geq 0$, we know that $k + 1 \geq 1$ and that A is nonempty. Let b be an element in A. We know from Lemma 7.3.2 that $|A-\{b\}| = (k+1)-1 = k$. Thus we can write $A - \{b\}$ as $A - \{b\} = \{a_1, a_2, \ldots, a_k\}$. Thus by the induction hypothesis, we know that $A - \{b\}$ has 2^k subsets. These are listed in the first column in Figure 7.6. Now form the union of each of these subsets with the set $\{b\}$ to form another list of subsets of A. These new subsets are listed in the second column of Figure 7.6.

All of the sets listed in Figure 7.6 are subsets of A. There are 2^k distinct subsets X of A listed in the first column of Figure 7.6. This implies that the list of 2^k subsets of the form $X \cup \{b\}$, for $X \subseteq A - \{b\}$, are also distinct. Furthermore, any subset X_1 of $A-\{b\}$ will not equal a subset of the form $X_2 \cup \{b\}$ for $X_2 \subseteq A-\{b\}$ as $b \notin X_1$ and $b \in X_2 \cup \{b\}$. In other words, no set in the first column of Figure 7.6 equals a set listed in the second column of Figure 7.6. Putting all this together means that this process has created $2^k + 2^k = 2^{k+1}$ subsets of A.

Lastly, we must show that every subset of A is of the form X_1, where $X_1 \subseteq A - \{b\}$, or of the form $X_2 \cup \{b\}$, where $X_2 \subseteq A - \{b\}$. To do this, let Y be a subset of A. We have two cases to examine.

Case (1) $b \notin Y$. In this case, $Y \subseteq A - \{b\}$ and Y is one of the 2^k subsets of $A - \{b\}$ listed in the first column of Figure 7.6.

Case (2) $b \in Y$. For this case, consider the set $Y - \{b\}$. Then $Y - \{b\}$ is a subset of $A - \{b\}$ and is again listed in the first column of Figure 7.6. This means that $(Y - \{b\}) \cup \{b\} = Y$ is listed in the second column of Figure 7.6.

Since every subset of A has been created in this process and 2^{k+1} subsets of Y were created, we have that $|\mathcal{P}(A)| = 2^{k+1}$. $\qquad\square$

Now to prove two results concerning infinite sets. To begin, recall that a set is infinite when it is not finite.

Lemma 7.3.6. *The set $\mathbb{N}$ is infinite.*

Proof. Suppose that $\mathbb{N}$ is finite. Since $\mathbb{N} \neq \emptyset$, it has the same cardinality as the set $\{1, 2, \ldots, n\}$, for some $n \in \mathbb{N}$, and there exists a bijection $f : \{1, 2, \ldots, n\} \to \mathbb{N}$. The set of natural numbers $A = \{f(1), f(2), \ldots, f(n)\}$ is a subset of $\mathbb{N}$. Using the ordering of $\mathbb{N}$ established by the Well-Ordering Principle (Axiom 4.1.1), the set of natural numbers A will have a largest element. Without loss of generality, let $f(n)$ be the largest natural number in A. In other words, $f(i) \leq f(n)$ for $1 \leq i \leq n$. Consider the natural number $f(n) + 1$. Since $f(i) < f(n) + 1$, for $1 \leq i \leq n$, it is in $\mathbb{N}$ with no element in the set $\{1, 2, \ldots, n\}$ mapping to it. Thus f is not onto. This contradiction means that $\mathbb{N}$ is infinite. $\quad\square$

Lemma 7.3.7. *Let A and B be sets such that $A \subseteq B$. If A is infinite, then B is infinite.*

The proof is left as an exercise. Using Lemma 7.3.7 and the fact that $\mathbb{N} \subseteq \mathbb{Z}, \mathbb{N} \subseteq \mathbb{Q}$, and $\mathbb{N} \subseteq \mathbb{R}$, it follows directly that $\mathbb{Z}, \mathbb{Q}$, and $\mathbb{R}$ are also infinite.

Exercises 7.3

(1) Let A and B be finite sets such that $|A| = n$ and $|B| = m$. Determine the following cardinalities.

(a) $|A \times B \times A|$

(b) $|A \times \mathcal{P}(B)|$

(c) $|\mathcal{P}(A) \times \mathcal{P}(B)|$

(d) $|\mathcal{P}(A \times B)|$

(e) $|\mathcal{P}(\mathcal{P}(A))|$

(f) $|\emptyset \times \mathcal{P}(A)|$

(g) $|\mathcal{P}(\emptyset) \times \mathcal{P}(\mathcal{P}(\emptyset))|$

(h) $|A \times \mathcal{P}(A) \times \mathcal{P}(\mathcal{P}(A))|$

(2) Find the cardinality of the following sets A_i, for $i \in \{1,2,3,4\}$, by finding a bijection $f : \{1,2,\ldots,n\} \to A_i$ for the appropriate $n \in \mathbb{N}$.

(a) $A_1 = \{2,4,6,\ldots,100\}$

(b) $A_2 = \{3,8,13,18,\ldots,103\}$

(c) $A_3 = \{\frac{1}{64}, \frac{1}{32}, \frac{1}{16}, \ldots, 64\}$.

(d) $A_4 = \{1, \frac{1}{3}, \frac{1}{6}, \frac{1}{10}, \ldots, \frac{1}{55}\}$

(3) Prove Lemma 7.3.2 (ii): given a set A with $|A| = n$, then for any element $d \notin A$, $|A \cup \{d\}| = n + 1$.

(4) Prove Theorem 7.3.3: for sets A and B, with $A \subseteq B$ and B finite, then A is finite and $|A| \leq |B|$.

(5) Let A and B be finite sets and consider the function $f : A \to B$.

(a) Prove that if f is one-to-one, then $|A| \leq |B|$.

(b) Prove that if f is onto, then $|A| \geq |B|$.

(6) Prove that if A and B are finite, then $A \cup B$ is finite.

(7) Prove that if $A_1, A_2, \ldots, A_n$, for $n \in \mathbb{N}$, are finite, then $A_1 \cup A_2 \cup \cdots \cup A_n$ is finite.

(8) Prove that if $A_1, A_2, \ldots, A_n$, for $n \in \mathbb{N}$, are finite, then $|A_1 \times A_2 \times \cdots \times A_n| = |A_1| \cdot |A_2| \cdots |A_n|$.

(9) Prove Lemma 7.3.7: for sets A and B such that $A \subseteq B$, if A is infinite, then B is infinite.

7.4 COUNTABLY INFINITE SETS

Definition 7.4.1. *Let A be a set.*

(i) *The set A is **countable** or **denumerable** if and only if it is finite or has the same cardinality as $\mathbb{N}$.*

(ii) *The set A is **countably infinite** if and only if it is countable and not finite.*

Recall that in Example 7.2.2 (ii) we showed that the set $O^+ = \{1, 3, 5, 7, 9, \ldots\}$, the set of positive odd integers, has the same cardinality as $\mathbb{N}$. Thus O^+ is countably infinite. Notice how we can list the elements of O^+ in an ordered list as $1, 3, 5, 7, 9, \ldots$ This is not by chance.

Suppose set A is countably infinite. Then there exists a bijection $f : \mathbb{N} \to A$. Since f is onto, for every element $a \in A$, there exists an $i \in \mathbb{N}$ such that $f(i) = a$. By denoting $a = f(i) = a_i$, this implies that every element of A can be written in the form a_j, for $j \in \mathbb{N}$. Furthermore, the representation is unique. If $a_i = a_j$, then we have $f(i) = f(j)$. But given that f is one-to-one, this implies $i = j$. This means we can use the ordering of $\mathbb{N} = \{1, 2, 3, 4, 5, 6, \ldots\}$ to write the elements of A in the ordered list

$$a_1, a_2, a_3, a_4, a_5, a_6, \ldots$$

This motivates the following result.

Theorem 7.4.2. *A set A is countably infinite if and only if the elements from A can be written in an infinite ordered list $a_1, a_2, a_3, a_4, \cdots$*

This observation is central to a number of results given in this section. We start with the following result.

Theorem 7.4.3. *Any subset of a countable set is countable.*

Proof. Let A be a countable set, and let B be a subset of A. If A is finite, then B is finite by Lemma 7.3.3 and the result follows. Proceed assuming that A is an infinite countable set. If B is finite, then by definition it is countable. Thus assume B is infinite.

Since A is countable, we can write $A = \{a_1, a_2, a_3, a_4, \ldots\} = \{a_i : i \in \mathbb{N}\}$. Let $I = \{i \in \mathbb{N} : a_i \in B\}$. We will now build a set C using induction. By the Well-Ordering Principle (Axiom 4.1.1), there exists a smallest natural number i_1 in I. This means $a_{i_1} \in B$. Let $b_1 = a_{i_1}$ and start building C by letting $C = \{b_1\}$. Now consider the set $I - \{i_1\}$. Since I is infinite, $I - \{i_1\}$ is nonempty. By the Well-Ordering Principle (Axiom 4.1.1), there exists a smallest element i_2 in $I - \{i_1\}$. It follows that $a_{i_2} \in B$. Let $b_2 = a_{i_2}$ and add it to C to get $C = \{b_1, b_2\}$. These two elements are distinct. If not, then $b_1 = b_2$, implying that $a_{i_1} = a_{i_2}$ with $i_1 \neq i_2$. This contradiction implies $b_1 \neq b_2$.

Now assume that we have completed this process k-times, with $k \geq 2$.

In other words, i_k is the smallest natural number in $I - \{i_1, i_2, \ldots, i_{k-1}\}$, with $b_k = a_{i_k} \in B$, creating the set $C = \{b_1, b_2, \ldots, b_k\}$ of distinct element from B. Now consider the infinite set $I - \{i_1, i_2, \ldots, i_k\}$. Let i_{k+1} be the smallest natural number in $I - \{i_1, i_2, \ldots, i_k\}$. Thus we have $a_{i_{k+1}} \in B$. Let $b_{k+1} = a_{i_{k+1}}$ and add it to C, creating $C = \{b_1, b_2, \ldots, b_k, b_{k+1}\}$. If $b_{k+1} = b_j$, for $1 \le j \le k$, then $a_{i_{k+1}} = a_{i_j}$ with $j \le k + 1$. This contradiction means $b_{k+1} \ne b_j$, for $1 \le j \le k$.

We have now created an infinite set $C = \{b_1, b_2, b_3, b_4, \ldots\}$ of distinct elements from B which is countable. If we can show $B = C$, we are done. It follows that $C \subseteq B$ due to the method used to create C. To show that $B \subseteq C$, let $b \in B$. Since B is a subset of A, we know that $b = a_n$ for some $n \in \mathbb{N}$. This means that $n \in I$. If $n = i_1$, $b = a_{i_1} = b_1$ and $b \in C$. Assume that $n \ne i_1$. Let $J = \{i_1, i_2, \ldots, i_t\}$ where $i_j \in I$ and $i_j < n$, for $1 \le j \le t$. Now consider the set $I - J$. The smallest natural number in $I - J$, which exists by the Well-Ordering Principle, is i_{t+1}. Given that $i_{t+1} \not< n$, this forces $i_{t+1} = n$. This means $b = a_{i_{t+1}} = b_{t+1}$ and that $b \in C$. This we have $B = C$ and the subset B of A is countable. □

As an exercise, you will show that $\mathbb{N}$ is countably infinite. So what other sets are countably infinite? Is $\mathbb{Z}$ countable infinite? You can write the elements of $\mathbb{Z}$ in an ordered list, as shown below, indicating by Theorem 7.4.2 that the answer is yes.

$\mathbb{N}$	1	2	3	4	5	6	7	8	9	$\cdots$
$\mathbb{Z}$	0	1	-1	2	-2	3	-3	4	-4	$\cdots$

Theorem 7.4.4. *The integers $\mathbb{Z}$ are countably infinite.*

The function $f : \mathbb{N} \to \mathbb{Z}$, defined by $f(n) = \frac{1 + (-1)^n (2n-1)}{4}$ is a bijection that also shows that $\mathbb{Z}$ is countably infinite. So let's now move on to the rational numbers $\mathbb{Q}$.

Theorem 7.4.5. *The rational numbers $\mathbb{Q}$ are countably infinite.*

Proof. We know by Lemma 7.3.7 that $\mathbb{Q}$ is infinite. To show it is countable, consider the matrix of rational numbers presented in Figure 7.7. Its columns are labeled in the order of the integers $\mathbb{Z}$ established earlier. The rows are labeled in the order of the natural numbers $\mathbb{N}$. Let $i \in \mathbb{N}$ and $j \in \mathbb{Z}$. The entry in row i and column j is the rational number $\frac{j}{i}$.

We can use this matrix to create an ordered list of the rational numbers. Start in the upper left of the matrix with the entry $\frac{0}{1} = 0$. Now follow the route, outlined in Figure 7.7, that snakes through the matrix.

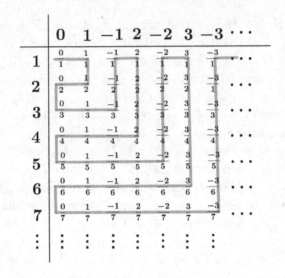

Figure 7.7 Matrix of $\mathbb{Q}$.

Each time you arrive at a rational number that is in lowest terms and that you have not yet written in the ordered list, add it to the ordered list. This creates the following ordered list of the rational numbers.

N	1	2	3	4	5	6	7	8	9	10	11	12	13	$\cdots$
$\mathbb{Q}$	0	1	$\frac{1}{2}$	$\frac{1}{3}$	$\frac{-1}{3}$	$\frac{-1}{2}$	-1	2	$\frac{2}{3}$	$\frac{-1}{4}$	$\frac{1}{4}$	$\frac{1}{5}$	$\frac{-1}{5}$	$\cdots$

Since every $q \in \mathbb{Q}$ will appear in lowest terms somewhere in this matrix, we have that $\mathbb{Q}$ is countably infinite by Theorem 7.4.2. □

The proof of Theorem 7.4.5 is sometimes referred to as the Snake Proof due to the method used to show the rational numbers can be arranged in an ordered list. This method motivates the following result, whose proof is very similar to the proof of Theorem 7.4.5.

Theorem 7.4.6. *If A and B are countable, then $A \times B$ is countable.*

Proof. There are three cases to examine.

   Case (1) A and B are finite. This case follows from Lemma 7.3.4.

   Case (2) One set is finite and the other is infinite. Assume without loss of generality that A is finite and that B is infinite. Then $A = \{a_1, a_2, \ldots, a_n\}$ and $B = \{b_1, b_2, b_3, \ldots\} = \{b_j : j \in \mathbb{N}\}$. Thus each element in $A \times B$ can be expressed in the form (a_i, b_j), where $i \in \mathbb{N}$, with $1 \leq i \leq n$, and $j \in \mathbb{N}$. Define a mapping $f : A \times B \to \mathbb{N}$ by

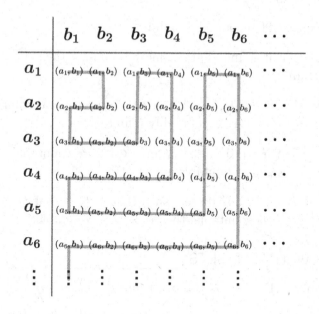

Figure 7.8 Matrix of $A \times B$.

$f(a_i, b_j) = (j-1)n + i$ for $(a_i, b_j) \in A \times B$. The proof that f is a bijection is almost identical to the one used in the proof of Lemma 7.3.4 and is omitted. This shows that $A \times B$ is countable.

Case (3) Both A and B are infinite. We know we can write $A = \{a_1, a_2, a_3, \ldots\} = \{a_i : i \in \mathbb{N}\}$ and $B = \{b_1, b_2, b_3, \ldots\} = \{b_j : j \in \mathbb{N}\}$. Consider the matrix, shown in Figure 7.8, where the ith row is labeled along the left by a_i and the jth column is labeled on the top by b_j. Each (i, j) entry in the matrix is the ordered pair (a_i, b_j).

We can now create an ordered list of the elements in $A \times B$. Start in the upper left of the matrix with the entry (a_1, b_1). Now follow the route, outlined in Figure 7.8, that snakes through the matrix. This creates the ordered list of the elements in $A \times B$ as documented below.

$\mathbb{N}$	1	2	3	4	5	6	7	$\cdots$
$A \times B$	(a_1, b_1)	(a_1, b_2)	(a_2, b_2)	(a_2, b_1)	(a_3, b_1)	(a_3, b_2)	(a_3, b_3)	$\cdots$

Thus by Theorem 7.4.2, $A \times B$ is countably infinite. □

Exercises 7.4

(1) Show that the function $f : \mathbb{N} \to \mathbb{Z}$, defined by $f(n) = \frac{1 + (-1)^n (2n-1)}{4}$ is one-to-one and onto.

(2) Prove that $\mathbb{N}$ is countably infinite.

(3) Prove that if A and B are countably infinite, then $A \cup B$ is countable infinite.

(4) Prove that if $A_1, A_2, \ldots, A_n$, for $n \in \mathbb{N}$, are countably infinite, then $A_1 \cup A_2 \cup \cdots \cup A_n$ is countably infinite.

(5) Prove that if $A_1, A_2, \ldots, A_n$, for $n \in \mathbb{N}$, are countable, then $A_1 \times A_2 \times \cdots \times A_n$ is countable.

(6) Let A, B, C, and D be sets such that $C \subseteq A$ and $D \subseteq B$. Prove that $A \times B$ is countable, then $C \times D$ is countable.

7.5 UNCOUNTABLE SETS

Definition 7.5.1. *A set is **uncountable** if and only if it is not countable.*

At this point, given that we have shown that $\mathbb{N}, \mathbb{Z}$, and $\mathbb{Q}$ are all countable, it appears that it might be that all sets are countable. However, this is not true. It turns out that the real numbers $\mathbb{R}$ are uncountable. To show this, we need the following lemma.

Lemma 7.5.2. *Let A and B be sets such that $A \subseteq B$. If A is uncountable, then B is uncountable.*

Proof. Suppose that B is countable. Then by Theorem 7.4.3, we know that A is countable. This contradiction implies that B is uncountable. □

Theorem 7.5.3. *The interval $(0,1)$ is uncountable.*

Proof. Assume that the interval $(0,1)$ is countable. This implies that we can write the elements of $(0,1)$ as $\{r_1, r_2, r_3, r_4, \ldots, r_n, \ldots\}$. Given that $0 < r_i < 1$, for $i \in \mathbb{N}$, each has a decimal expansion of the form $r_i = 0.a_{i,1}a_{i,2}a_{i,3}a_{i,4}\cdots$ where each $a_{i,j}$, for $j \in \mathbb{N}$ is a digit (i.e., $a_{i,j} \in \{0,1,2,3,4,5,6,7,8,9\}$). Now write the elements of $(0,1)$ in a vertical ordered list as shown in Figure 7.9.

Define the real number $r = 0.b_1b_2b_3b_4\cdots$ where each digit b_k, for $k \in \mathbb{N}$, is defined by

$$b_k = \begin{cases} 3 & \text{if } a_{k,k} \neq 3 \\ 4 & \text{if } a_{k,k} = 3 \end{cases}.$$

$$r_1 = 0.a_{1,1}\,a_{1,2}\,a_{1,3}\,a_{1,4}\cdots$$
$$r_2 = 0.a_{2,1}\,a_{2,2}\,a_{2,3}\,a_{2,4}\cdots$$
$$r_3 = 0.a_{3,1}\,a_{3,2}\,a_{3,3}\,a_{3,4}\cdots$$
$$r_1 = 0.a_{4,1}\,a_{4,2}\,a_{4,3}\,a_{4,4}\cdots$$
$$\vdots$$
$$r_n = 0.a_{n,1}\,a_{n,2}\,a_{n,3}\,a_{n,4}\cdots a_{n,n}\cdots$$
$$\vdots$$

Figure 7.9 Elements in $(0, 1)$.

The number r is in the interval $(0, 1)$ and it does not equal any of the numbers r_i, for $i \in \mathbb{N}$. To show this, suppose for some $j \in \mathbb{N}$ that $b = a_j$. This implies that

$$0.b_1 b_2 b_3 b_4 \cdots = 0.a_{j,1} a_{j,2} a_{j,3} a_{j,4} \cdots.$$

For these two numbers to be equal, we must have $b_1 = a_{j,1}, b_2 = a_{j,2}, b_3 = a_{j,3}$, and so on. More generally, $b_k = a_{j,k}$ for each $k \in \mathbb{N}$. This implies that $b_j = a_{j,j}$. But by the definition of the number b, if $a_{j,j} = 3$, then $b_j = 4$, and if $a_{j,j} \neq 3$, then $b_j = 3$. This means that $b_j \neq a_{j,j}$ and that $b \neq a_j$. This is a contradiction as $r \in (0, 1)$ is not in the list of elements in $(0, 1)$. Consequently the interval $(0, 1)$ is uncountable. □

The method used to prove this result, as visualized in Figure 7.9, it is often referred to as the Diagonal Proof. We can now prove the result we have been waiting for.

Theorem 7.5.4. *The set of real numbers $\mathbb{R}$ is uncountable.*

Proof. By Theorem 7.2.4, we know that the interval $(0, 1)$ has the same cardinality as the real numbers. Since the interval $(0, 1)$ is uncountable, as proven in Theorem 7.5.3, the real numbers are uncountable. □

We finish this short section with two important results concerning power sets. The first is motivated by Theorem 7.3.5 where it was observed that $|A| \neq |\mathcal{P}(A)| = 2^{|A|}$ for a finite set A.

Theorem 7.5.5. *For A a set, A and $\mathcal{P}(A)$ do not have the same cardinality.*

Proof. First suppose that $A = \emptyset$. Then $|A| = 0$, with $|\mathcal{P}(A)| = |\{\emptyset\}| = 1$, and A and $\mathcal{P}(A)$ do not have the same cardinality.

Proceed assuming that $A \neq \emptyset$ and that A and $\mathcal{P}(A)$ have the same cardinality. Thus there exists a bijective function $f : A \to \mathcal{P}(A)$. Let $B = \{a \in A : a \notin f(a)\}$. Since f is onto, there exists an element $b \in A$ such that $f(b) = B$. We now have two cases to examine:

Case (1) $b \in B$. This implies that $b \in f(b)$. Thus by the definition of B, we have that $b \notin f(b) = B$, a contradiction.

Case (2) $b \notin B$. Therefore, we have $b \notin f(b)$, which implies by the definition of B that $b \in B$, another contradiction.

Consequently, we have a contradiction, yielding that A and $\mathcal{P}(A)$ do not have the same cardinality. $\qquad\square$

Theorem 7.5.6. *The set $\mathcal{P}(\mathbb{N})$ is uncountable.*

Proof. By Theorem 7.5.5, we know that $\mathbb{N}$ and $\mathcal{P}(\mathbb{N})$ do not have the same cardinality. To show that $\mathcal{P}(\mathbb{N})$ is uncountable, all we need to show is that $\mathcal{P}(\mathbb{N})$ is not finite. Consider the set $A = \{\{1\}, \{2\}, \{3\}, \{4\}, \ldots\}$. Define the mapping $f : \mathbb{N} \to A$ by $f(n) = \{n\}$ for $n \in \mathbb{N}$. It follows directly that f is a bijection. Thus A has the same cardinality as $\mathbb{N}$. Thus by Lemma 7.3.6, we know that A is infinite. Given that $A \subseteq \mathcal{P}(N)$ and A is infinite, we know by Lemma 7.3.7 that $\mathcal{P}(\mathbb{N})$ is infinite. This means that the set $\mathcal{P}(\mathbb{N})$ is uncountable. $\qquad\square$

Exercises 7.5

(1) Let A and B be uncountable sets. Prove that $A \times B$ and $A \cup B$ are uncountable.

(2) Show that if A and B are uncountable sets, then $A \cap B$ is not necessarily uncountable.

(3) Let A be a nonempty set. Prove that $A \times \mathbb{R}$ is uncountable.

(4) Prove that $\mathcal{P}(\mathbb{Z})$ is uncountable.

7.6 COMPARING CARDINALITIES

When you are given two finite sets A and B, it is fairly straightforward to compare their cardinalities or orders by using the ordering of the natural numbers. Let $|A| = n$ and $|B| = m$ with $n, m \in \mathbb{N} \cup \{0\}$. As integers, if $n < m$, we can then say $|A| < |B|$ or if $n = m$, then $|A| = |B|$. But how do we compare cardinalities of infinite sets? The answer is motivated by observations we made concerning one-to-one and onto functions in Chapter 6.

Definition 7.6.1. *Let A and B be sets.*

  (i) *If there exists a bijective function from A to B, then $|A| = |B|$.*

 (ii) *If there exists an one-to-one function from A to B, but no onto function from A to B, then $|A| < |B|$.*

(iii) *If $|A| < |B|$ or $|A| = |B|$, then $|A| \leq |B|$.*

For a quick example, we will show that $|\mathbb{N}| < |\mathbb{R}|$. Define the function $f : \mathbb{N} \to \mathbb{R}$ by $f(n) = n$ for $n \in \mathbb{N}$. It is simple to show that f is one-to-one. Now consider the element $\sqrt{2} \in \mathbb{R}$. We know by Theorem 2.7.2 that $\sqrt{2}$ is not rational and in particular is not an integer. Thus there is no $m \in \mathbb{N}$ such that $f(m) = \sqrt{2}$ and f is not onto. We know by Theorem 7.5.4 that $\mathbb{R}$ is uncountable. This means there does not exist a bijective function from $\mathbb{N}$ to $\mathbb{R}$. Thus we have $|\mathbb{N}| < |\mathbb{R}|$.

We proved in Theorem 7.5.5 for any given set A that A and $\mathcal{P}(A)$ do not have the same cardinality. We now prove a stronger result. It is due to Georg Cantor (1845-1918).

Theorem 7.6.2. *If A is a set, then $|A| < |\mathcal{P}(A)|$.*

Proof. First consider the case that A is finite. Then we have $|A| = n$, with $n \in \mathbb{N} \cup \{0\}$. By Theorem 7.3.5, we know that $|\mathcal{P}(A)| = 2^n$. Given that $n < 2^n$ by Lemma 4.2.10, we get $|A| < |\mathcal{P}(A)|$.

Now consider the case that A is infinite. Define a function $f : A \to \mathcal{P}(A)$ by $f(a) = \{a\}$ for all $a \in A$. Let $a, b \in A$ and assume that $f(a) = f(b)$. This implies that $\{a\} = \{b\}$ or that $a = b$. Thus the function f is one-to-one. Now consider $\emptyset \in \mathcal{P}(A)$. If there existed an element $x \in A$ such that $f(x) = \emptyset$, this would imply that $f(x) = \{x\} = \emptyset$, which is a contradiction. Thus f is not onto.

We know by Theorem 7.5.5 that A and $\mathcal{P}(A)$ do not have the same cardinality. Thus there does not exist a bijective function from A to $\mathcal{P}(A)$. This implies that $|A| < |\mathcal{P}(A)|$. $\qquad\square$

The cardinality of the natural numbers $\mathbb{N}$ is often denoted by $\aleph_0$, which we read as "aleph-naught." This in some sense is a measure of the infinity represented by the infinite set $\mathbb{N}$. But given that $|\mathbb{N}| < |\mathcal{P}(\mathbb{N})|$ by Theorem 7.6.2, we have $\aleph_0 < |\mathcal{P}(\mathbb{N})|$. This implies that there are larger measures of infinity. While not proven here, it should also be noted that $\mathcal{P}(\mathbb{N})$ has the same cardinality as $\mathbb{R}$.

Given that $\mathcal{P}(\mathbb{N})$ is also an infinite set, we have that $\aleph_0 < |\mathcal{P}(\mathbb{N})|$. The measure of infinity given by $|\mathcal{P}(\mathbb{N})|$ is sometimes referred to as $\aleph_1$. This motivates the following question: Does there exist a set A such that $\aleph_0 < |A| < \aleph_1$? Currently, this question is unresolved. The general feeling about the answer to this question is stated here.

Continuum Hypothesis. *There is no set A such that $\aleph_0 < |A| < \aleph_1$.*

Given that $\mathcal{P}(\mathbb{N})$ is a set, we can form the set $\mathcal{P}(\mathcal{P}(\mathbb{N}))$. We know by Theorem 7.6.2 that $|\mathcal{P}(\mathbb{N})| < |\mathcal{P}(\mathcal{P}(\mathbb{N}))|$. This means that there is a third measure of infinity. But why stop there? Applying Theorem 7.6.2 repeatedly, we get that

$$|\mathbb{N}| < |\mathcal{P}(\mathbb{N})| < |\mathcal{P}(\mathcal{P}(\mathbb{N}))| < |\mathcal{P}(\mathcal{P}(\mathcal{P}(\mathbb{N})))| < |\mathcal{P}(\mathcal{P}(\mathcal{P}(\mathcal{P}(\mathbb{N}))))| < \cdots$$

This means that there are countably many measures of infinity. If we define, for each $n \in \mathbb{N}$,

$$\aleph_n = \underbrace{\mathcal{P}(\mathcal{P}(\cdots \mathcal{P}(\mathbb{N}) \cdots))}_{n-\text{times}},$$

the observation above can be restated as

$$\aleph_0 < \aleph_1 < \aleph_2 < \aleph_3 < \cdots < \aleph_n < \cdots .$$

This means there is no largest infinite cardinality.

We end with one final comment. Cantor's result (Theorem 7.6.2) leads to another paradox in naive set theory discovered by him called Cantor's Paradox. Let X be the set of all sets. By Theorem 7.6.2, we have that $|X| < |\mathcal{P}(X)|$. However, given that $\mathcal{P}(X)$ is also a set, it follows that $\mathcal{P}(X) \subseteq X$. This implies that $|\mathcal{P}(X)| \leq |X|$. This paradox, along with Russell's Paradox described in Section 3.6, motivated the development of axiomatic set theory.

Exercises 7.6

(1) Prove that $|\mathbb{Q}| < |\mathbb{R}|$.

(2) Let A and B be nonempty sets. Prove that $|A| \leq |A \times B|$.

(3) Let A and B be sets. Prove that $|A| \leq |A \cup B|$.

(4) Let A and B be sets. Prove that $|A \cap B| \leq |A|$.

CHAPTER 8

Conclusion

I hope that by reading this book and doing the exercises, you have gained new tools for your journey into mathematics. The concepts introduced here have prepared you to tackle the upper-level mathematics courses that you will now take. You will learn numerous new topics in these courses, and many of the concepts introduced in this text will be investigated in more detail in them. Your mathematical journey has only started and learning never ends.

In addition to developing competencies, I also hope that this book has reinforced your appreciation for the power and beauty of mathematics. Mathematics is a rich subject area with applications in almost every discipline. But there is also a simplicity, order, and flow to mathematics that is aesthetically pleasing. The more you learn, the more you will appreciate its poetry.

I wish you all of the best as you continue to travel down the road of mathematics.

CHAPTER 9

Hints and Solutions to Selected Exercises

Chapter 1 Exercises

Section 1.2

(1) (a) This is a true statement.   (e) This is a false statement.

(c) This is not a statement.   (f) This is not a statement.

(2) (a) This statement is false.   (g) This statement is false.

(c) This statement is true.   (i) This statement is true.

(e) This statement is false.   (k) This statement is true.

Section 1.3

(1) (b) The statement $(P \vee Q) \to \sim (P \wedge Q)$ is a contingency.

P	Q	$P \vee Q$	$P \wedge Q$	$\sim (P \wedge Q)$	$(P \vee Q) \to \sim (P \wedge Q)$
T	T	T	T	F	F
T	F	T	F	T	T
F	T	T	F	T	T
F	F	F	F	T	T

(2) (a) The statement $\sim Q \wedge (P \to Q) \to \sim P$ is a tautology. In the truth table, S_1 denotes $\sim Q \wedge (P \to Q)$.

P	Q	$\sim Q$	$P \to Q$	S_1	$\sim P$	$S_1 \to \sim P$
T	T	F	T	F	F	T
T	F	T	F	F	F	T
F	T	F	T	F	T	T
F	F	T	T	T	T	T

(4) Two statements S_1 and S_2 satisfy $S_1 \equiv S_2$ if $S_1 \leftrightarrow S_2$ is a tautology.

(a)

P	$\sim P$	$\sim (\sim P)$	$P \leftrightarrow \sim (\sim P)$
T	F	T	T
F	T	F	T

(c) Let S_1 denote $(P \wedge Q) \wedge R$ and S_2 denote $P \wedge (Q \wedge R)$.

P	Q	R	$P \wedge Q$	S_1	$Q \wedge R$	S_2	$S_1 \leftrightarrow S_2$
T	T	T	T	T	T	T	T
T	T	F	T	F	F	F	T
T	F	T	F	F	F	F	T
T	F	F	F	F	F	F	T
F	T	T	F	F	T	F	T
F	T	F	F	F	F	F	T
F	F	T	F	F	F	F	T
F	F	F	F	F	F	F	T

(e) Let S_1 denote $\sim (P \wedge Q)$ and S_2 denote $\sim P \vee \sim Q$.

P	Q	$P \wedge Q$	S_1	$\sim P$	$\sim Q$	S_2	$S_1 \leftrightarrow S_2$
T	T	T	F	F	F	F	T
T	F	F	T	F	T	T	T
F	T	F	T	T	F	T	T
F	F	F	T	T	T	T	T

Section 1.4

(1) (a) This statement is true. For all real numbers x, we have $e^x > 0$. Therefore, it follows that $e^x + 1 > 0 + 1$ or $e^x > 1$.

(d) This statement is false. The only values of z that satisfy $z^2 = 3$ are $z = -\sqrt{3}$ and $z = \sqrt{3}$, neither of which are integers.

(g) This statement is false. Consider the integer $z = -3$. Given that $n \geq 1$ for any $n \in \mathbb{N}$, no natural number n satisfies $n < -3$.

(j) This statement is true. Let $r = \frac{3}{2}$ and $s = \frac{1}{2}$.

(m) This statement is false. Consider the values of $x = -2$ and $y = 1$. Here we have that the statement $x^2 > y^2$ is true given that $(-2)^2 > 1^2$, but the statement $x > y$ is false since $-2 \not> 1$.

Section 1.5

(1) (a) The integer 32 is odd or not a power of 2.

(c) For all real numbers x, $x^2 \leq 1$ or $x \geq 0$.

(e) There exist integers n and m, such that $n < 0$, $m < 0$, and $nm \leq 0$.

Chapter 2 Exercises

Section 2.4

(1) *Proof.* Given that n is even and m is odd, there exist integers k and l such that $n = 2k$ and $m = 2l + 1$. This implies

$$n + m = 2k + 2l + 1 = 2(k + l) + 1.$$

Since $k + l$ is an integer, $n + m$ is odd. $\square$

(4) *Proof.* Given that $3n + 2$ is odd, there exists an integer k such that $3n + 2 = 2k + 1$. Adding $2n$ to both sides of this equation results in $5n + 2 = 2k + 2n + 1$. Adding 5 to both sides of this equation yields $5n + 7 = 2k + 2n + 6$. This implies that $5n + 7 = 2(k + n + 3)$. Since $k + n + 3$ is an integer, the number $5n + 7$ is even. $\square$

(5) (b) *Proof.* Given that $a|c$ and $b|d$, there exist integers k and l such that $c = ka$ and $d = lb$. This implies

$$cd = (ka)(lb) = k(al)b = k(la)b = (lk)ab.$$

Since lk is an integer, we have $ab|cd$. $\square$

(6) (a) *Proof.* Since we are assuming that $a \equiv b \pmod{n}$, it follows that $n|(a-b)$. Thus there exists an integer k such that $a-b = kn$. We now have

$$(a + c) - (b + c) = a + c - b - c = a - b = kn.$$

Since k is an integer, we have $n|[(a + c) - (b + c)]$. This implies that $a + c \equiv b + c \pmod{n}$. $\square$

(d) *Proof.* Given that $a \equiv b \pmod{n}$ and $c \equiv d \pmod{n}$, it follows that $n|(a-b)$ and $n|(c-d)$. Thus there exist integers k and l such that $a - b = kn$ and $c - d = ln$. Rewriting, we get that $a = b + kn$ and $c = d + ln$. We now have

$$ac - bd = (b + kn)(d + ln) - bd$$
$$= bd + (bl)n + (dk)n + (kln)n - bd$$
$$= (bl + dk + kln)n.$$

Since $bl + dk + kln$ is an integer, we have $n|(ac - bd)$. This shows that $ac \equiv bd \pmod{n}$. □

(9) *Proof.* First note that since $a \equiv b \pmod{n}$, we have $n|(a-b)$ or that $a - b = kn$ for some integer k. By the Division Algorithm (Theorem 2.3.8), there exist integers q_1, q_2, r_1, and r_2, with $0 \le r_1, r_2 \le n - 1$, such that $a = q_1 n + r_1$ and $b = q_2 n + r_2$. This implies

$$a - b = (q_1 n + r_1) - (q_2 n + r_2) = (q_1 - q_2)n + r_1 - r_2.$$

With $0 \le r_2 \le n-1$, we know $-n+1 \le -r_2 \le 0$. Since $0 \le r_1 \le n-1$, adding results in $-n+1 \le r_1 - r_2 \le n-1$. However, given that $a - b$ is a multiple of n, this forces $r_1 - r_2 = 0$ or that $r_1 = r_2$. Therefore, the integers a and b have the same remainder when divided by n. □

(12) *Proof.* Since r and s are rational, there exist integers a, b, c and d, with $b \ne 0$ and $d \ne 0$, such that $r = \frac{a}{b}$ and $s = \frac{c}{d}$. We now have

$$\frac{r + s}{2} = \frac{\frac{a}{b} + \frac{c}{d}}{2} = \frac{\frac{ad}{bd} + \frac{bc}{bd}}{2} = \frac{\frac{ad+bc}{bd}}{2} = \frac{ad + bc}{2bd},$$

where $ad + bc$ and $2bd$ are integers. Furthermore, given that $b \ne 0$ and $d \ne 0$, we know that $2bd \ne 0$. It follows that $\frac{r+s}{2}$ is rational. □

Section 2.5

(1) *Proof.* Assume that a is even. Then there exists an integer k such that $a = 2k$. This implies

$$5a + 7 = 5(2k) + 7 = 10k + 6 + 1 = 2(5k + 3) + 1,$$

where $5k + 3$ is an integer. This means $5a + 7$ is odd. □

(4) *Proof.* Assume we have $5|n$. Then there exists an integer k such that $n = 5k$. This implies

$$n^2 = (5k)^2 = 25k^2 = 5(5k^2),$$

where $5k^2$ is an integer. This means $5|n^2$. □

(7) *Proof.* Suppose $a \equiv 1 \pmod 3$. This means $3|(a-1)$ and that there exists an integer k such that $a - 1 = 3k$ or $a = 3k + 1$. Now rewrite $(a+1)^2 - 1$ to obtain

$$(a+1)^2 - 1 = (3k+1+1)^2 - 1 = 9k^2 + 12k + 4 - 1$$
$$= 9k^2 + 12k + 3 = 3(3k^2 + 4k + 1),$$

where $3k^2 + 4k + 1$ is an integer. This results in $3|((a+1)^2 - 1)$ and that $(a+1)^2 \equiv 1 \pmod 3$. □

(10) *Proof.* We are given that n is rational. Thus there exist integers a and b, with $b \neq 0$, such that $n = \frac{a}{b}$. Assume that m is not irrational or that m is rational. This means there exist integers c and d, with $d \neq 0$, such that $m = \frac{c}{d}$. This implies

$$n(m + 3) = \frac{a}{b}\left(\frac{c}{d} + 3\right) = \frac{a}{b}\left(\frac{c}{d} + \frac{3d}{d}\right) = \frac{a}{b}\left(\frac{c + 3d}{d}\right) = \frac{a(c + 3d)}{bd},$$

where $a(c + 3d)$ and bd are integers with $bd \neq 0$. Thus we have that $n(m + 3)$ is rational or that it is not irrational. □

Section 2.6

(1) *Proof.* Given that integers a and b are of the same parity, there are two cases to examine.

Case (1) a and b are both even. There exist integers k and l such that $a = 2k$ and $b = 2l$. This results in

$$a^2 + b^2 = (2k)^2 + (2l)^2 = 4k^2 + 4l^2 = 2(2k^2 + 2l^2),$$

where $2k^2 + 2l^2$ is an integer. This means $a^2 + b^2$ is even.

Case (2) a and b are both odd. There exist integers k and l such that $a = 2k + 1$ and $b = 2l + 1$. This results in

$$a^2 + b^2 = (2k+1)^2 + (2l+1)^2 = 4k^2 + 4k + 1 + 4l^2 + 4l + 1$$
$$= 4k^2 + 4l^2 + 4k + 4l + 2 = 2(2k^2 + 2l^2 + 2k + 2l + 1),$$

where $2k^2 + 2l^2 + 2k + 2l + 1$ is an integer. Thus $a^2 + b^2$ is even. □

(4) *Proof.* Given integers n and m, there are three cases to examine.

Case (1) n and m are both even. There exist integers k and l such that $n = 2k$ and $m = 2l$. This results in

$$n^2 + m^2 + 1 = (2k)^2 + (2l)^2 + 1 = 4k^2 + 4l^2 + 1 = 4(k^2 + l^2) + 1,$$

where $k^2 + l^2$ is an integer. Since $n^2 + m^2 + 1$ has a remainder of 1 when divided by 4, it is not a multiple of 4. This implies $4 \nmid (n^2 + m^2 + 1)$.

Case (2) n and m are both odd. There exist integers k and l such that $n = 2k + 1$ and $m = 2l + 1$. This results in

$$n^2 + m^2 + 1 = (2k + 1)^2 + (2l + 1)^2 + 1 = 4k^2 + 4l^2 + 4k + 4l + 3$$
$$= 4(k^2 + l^2 + k + l) + 3,$$

where $k^2 + l^2 + k + l$ is an integer. Since $n^2 + m^2 + 1$ has a remainder of 3 when divided by 4, this implies $4 \nmid (n^2 + m^2 + 1)$.

Case (3) n and m are of opposite parity. Without loss of generality, assume that n is even and that m is odd. Thus there exist integers k and l such that $n = 2k$ and $m = 2l + 1$. This results in

$$n^2 + m^2 + 1 = (2k)^2 + (2l + 1)^2 + 1 = 4k^2 + 4l^2 + 4l + 1 + 1$$
$$= 4k^2 + 4l^2 + 4l + 2 = 4(k^2 + l^2 + l) + 2,$$

where $k^2 + l^2 + l$ is an integer. Since $n^2 + m^2 + 1$ has a remainder of 2 when divided by 4, this implies $4 \nmid (n^2 + m^2 + 1)$. □

(7) HINT: By the Division Algorithm, it follows that $n = 3q + 0 = 3q, n = 3q + 1$, or $n = 3q + 2$ for integer q. Now examine the three cases.

(10) *Proof.* ($\Rightarrow$) Assume that n is even or m is even. Without loss of generality, assume that n is even. Thus there exists an integer k such that $n = 2k$. This implies

$$nm = (2k)m = 2(km),$$

where km is an integer. This implies that nm is even or that nm is not odd.

($\Leftarrow$) Assume that n and m are odd. Then there exist integers k and l such that $n = 2k + 1$ and $m = 2l + 1$. This results in

$$nm = (2k + 1)(2l + 1) = 4kl + 2k + 2l + 1 = 2(2kl + k + l) + 1,$$

where $2kl + k + l$ is an integer. This means that nm is odd. □

(14) HINT: First show that 1 and -1 divide 1. Then assume $1 = ab$, where $a, b \in \mathbb{Z}$ with a and b not equal to 1 or -1. Given that $a \neq 0$ and $b \neq 0$, this implies that $a, b < 0$ or $a, b > 0$. There are now two cases to examine.

Section 2.7

(1) HINT: Use Lemma 2.6.8.

(4) *Proof.* Suppose there exist an odd integer a and an even integer b such that $a^2 - b^2 = 0$. Since a is odd and b is even, there exist integers k and l such that $a = 2k + 1$ and $b = 2l$. Given $a^2 - b^2 = 0$, we have $(2k + 1)^2 - (2l)^2 = 0$ or that $4k^2 + 4k + 1 - 4l^2 = 0$. This implies $4l^2 - 4k^2 - 4k = 1$, yielding $4(l^2 - k^2 - k) = 1$ where $l^2 - k^2 - k$ is an integer. But this implies that 4 divides 1, which is a contradiction. Thus for all odd integers a and even integers b, we have $a^2 - b^2 \neq 0$. □

(7) HINT: Assume that $n^2 + 6n + 1$ is odd and that n is even, then find a contradiction.

(10) *Proof.* Assume that $a^3 \not\equiv b^3 \pmod 3$ and that $(a - b)^3 \equiv 0 \pmod 3$. Given that $(a-b)^3 \equiv 0 \pmod 3$, we know $3|(a-b)^3$. Thus there exists an integer k such that $(a-b)^3 = 3k$ or that $a^3 - 3a^2b + 3ab^2 - b^3 = 3k$. Rewriting, we have

$$a^3 - b^3 = 3k + 3a^2b - 3ab^2 = 3(k + a^2b - ab^2),$$

where $k + a^2b - ab^2$ is an integer. This implies $3|(a^3 - b^3)$ or that $a^3 \equiv b^3 \pmod 3$. This contradiction means that if $a^3 \not\equiv b^3 \pmod 3$, then $(a - b)^3 \not\equiv 0 \pmod 3$. □

Chapter 3 Exercises
Section 3.1

(1) (a) True. The number 2 is listed as an element in the set.

(c) False. While $\{2\}$ is an element in the set, the number 2 is not.

(e) True. The set $\{\{3\}\}$ has exactly one element $\{3\}$.

(g) True. The empty set is listed as an element in the set.

(2) (a) $\{-6, -3, -2, -1, 1, 2, 3, 6\}$

(c) $\{-1, 1\}$

(e) $\emptyset$

(g) $\{\ldots, -7, -3, 1, 5, 9, \ldots\}$

(i) $\{(1, 1, 1), (2, 1, 1), (1, 2, 1),$
$(1, 1, 2)\}$

(k) $\emptyset$

(m) $\{1, 4, 9, 16, 25, \ldots\}$

(3) (a) $\{n \in \mathbb{N} : 1 \leq n \leq 5\}$

(c) $\{6n : n \in \mathbb{Z}\}$

(e) $\{n \in \mathbb{N} : n \text{ is prime}\}$

(g) $\{5n - 3 : n \in \mathbb{Z}\}$

(i) $\{(n, n) : n \in \mathbb{Z}\}$

(k) $\{(a, b, c) \in \mathbb{N}^3 : a^2 + b^2 = c^2\}$

(5) (c) $\{(\emptyset, \emptyset, \emptyset), (\emptyset, \{\emptyset\}, \emptyset), (\emptyset, \emptyset, \{\emptyset\}), (\emptyset, \{\emptyset\}, \{\emptyset\}), (\emptyset, \emptyset, \{\emptyset, \{\emptyset\}\}),$
$(\emptyset, \{\emptyset\}, \{\emptyset, \{\emptyset\}\})\}$

Section 3.2

(1) (a) False. While $\{3\}$ is an element in the set, the number 3 is not.

(d) False. The set $\{3\}$ is an element of the set, not a subset.

(g) False. The empty set is a subset of the set.

(j) False. The set $\mathbb{R}$ is an element of the set, not a subset of it.

(m) True. Since $\{1\}$ is a subset of the set $\{1, 2, 3\}$, it is in the power set of $\{1, 2, 3\}$.

(p) True. Since $\{\emptyset\}$ is a subset of the set $\{\emptyset, 1\}$, it is in the power set of $\{\emptyset, 1\}$.

(5) *Proof.* ($\Rightarrow$) Suppose that $A \times B$ is not the empty set. Then there exists an element $(a, b) \in A \times B$. This means $a \in A$ and $b \in B$, implying that $A \neq \emptyset$ and $B \neq \emptyset$. This contradiction implies that $A \times B = \emptyset$.

($\Leftarrow$) Suppose that $A \neq \emptyset$ and $B \neq \emptyset$. Then there exists an element $a \in A$ and an element $b \in B$. This means (a, b) is an element in $A \times B$, a contradiction. Thus we have $A = \emptyset$ or $B = \emptyset$ □

(9) *Proof.* To show that $A \times C \subseteq B \times D$, let $(a, c) \in A \times C$. Given that $a \in A$, with $A \subseteq C$, it follows that $a \in C$. Similarly, since $c \in C$ with $C \subseteq D$, we have $c \in D$. This means $(a, c) \in B \times D$. □

(11) *Proof.* To show that $\mathcal{P}(A) \subseteq \mathcal{P}(B)$, let $X \in \mathcal{P}(A)$. This means that $X \subseteq A$. Given that $X \subseteq A$ and $A \subseteq B$, it follows by Lemma 3.2.2 that $X \subseteq B$. This means $X \in \mathcal{P}(B)$. □

Section 3.3

(1) (a) $\{11\}$

(c) $\{17, 19, 23, 29, 31\} = D$

(e) $\{3, 5, 7, 13, 17, 19, 29\}$

(g) $\{29\}$

(i) $\{2, 3, 5, 7, 13, 17, 19, 23, 31\}$

(6) *Proof.* To show $A \cup \emptyset \subseteq A$, let $a \in A \cup \emptyset$. Then $a \in A$ or $a \in \emptyset$. If $a \in A$, we are done. Now suppose that $a \in \emptyset$. This is a contradiction as the empty set contains no elements.

To show that $A \subseteq A \cup \emptyset$, let $a \in A$. It directly follows that $a \in A \cup \emptyset$. $\qquad\square$

(12) *Proof.* To show $A \times (B \cap C) \subseteq (A \times B) \cap (A \times C)$, let $(x, y) \in A \times (B \cap C)$. This means $x \in A$ and $y \in B \cap C$, yielding $y \in B$ and $y \in C$. Given that $x \in A$ and $y \in B$, it follows that $(x, y) \in A \times B$. Furthermore, since $x \in A$ and $y \in C$, we know $(x, y) \in A \times C$. This results in $(x, y) \in (A \times B) \cap (A \times C)$.

To show $(A \times B) \cap (A \times C) \subset A \times (B \cap C)$, let $(x, y) \in (A \times B) \cap (A \times C)$. This means $(x, y) \in A \times B$ and $(x, y) \in A \times C$. It then follows that $x \in A$ with $y \in B$ and $y \in C$. Thus we have $x \in A$ and $y \in B \cap C$, yielding $(x, y) \in A \times (B \cap C)$. $\qquad\square$

(14) *Proof.* To show $(A - B) \cup B \subseteq A$, let $x \in (A - B) \cup B$. This means $x \in A - B$ or $x \in B$. If $x \in A - B$, then $x \in A$ and $x \notin B$. Thus we have $x \in A$. If it is the case that $x \in B$, then given that $B \subseteq A$, we know that $x \in A$.

To show $A \subseteq (A - B) \cup B$, let $y \in A$. We know that $y \in B$ or $y \notin B$. If $y \in B$, then it follows that $y \in (A - B) \cup B$. Now suppose that $y \notin B$. Since $y \in A$ and $y \notin B$, we have $y \in A - B$. This means $y \in (A - B) \cup B$. $\qquad\square$

(21) *Proof.* To show $(A \cup B) - C \subseteq (A - C) \cup (B - C)$, let $x \in (A \cup B) - C$. It follows that $x \in A \cup B$ and that $x \notin C$. Since $x \in A$ or $x \in B$, first consider the possibility that $x \in A$. Since $x \in A$ and $x \notin C$, we have $x \in A - C$. This results in $x \in (A - C) \cup (B - C)$. Now consider the case that $x \in B$. Given that $x \in B$ and $x \notin C$, we have $x \in B - C$. This also results in $x \in (A - C) \cup (B - C)$.

To show $(A-C)\cup(B-C)\subseteq(A\cup B)-C$, let $y\in(A-C)\cup(B-C)$. We then have $y\in A-C$ or $y\in B-C$. If $y\in A-C$, then $y\in A$ and $y\notin C$. Since $y\in A$, we know $y\in A\cup B$. Therefore, it follows that $y\in A\cup B$ and $y\notin C$, implying that $y\in(A\cup B)-C$. Now if $y\in B-C$, then $y\in B$ and $y\notin C$. Since $y\in B$, we know $y\in A\cup B$. Therefore, it follows that $y\in A\cup B$ and $y\notin C$, implying that $y\in(A\cup B)-C$. □

(25) *Proof.* ($\Rightarrow$) Assume $A\cap B=A$. Let $x\in A$. This means $x\in A\cap B$. It then follows that $x\in B$.

($\Leftarrow$) Now assume that $A\subseteq B$. Let $x\in A\cap B$. This means $x\in A$ and $x\in B$. This implies $x\in A$. Now let $y\in A$. Given that $A\subseteq B$, it follows that $y\in B$. Since $y\in A$ and $y\in B$, we have $y\in A\cap B$. □

(27) HINT: The proof here is similar to the proof constructed for Exercise 12.

(31) *Proof.* To show $A-B\subseteq A\cap\overline{B}$, let $x\in A-B$. This means $x\in A$ and $x\notin B$. Given that $x\notin B$, it follows that $x\in\overline{B}$. Given that $x\in A$ and $x\in\overline{B}$, we know that $x\in A\cap\overline{B}$.

Proving that $A\cap\overline{B}\subseteq A-B$ is very similar. □

Section 3.4

(1) $\bigcup_{i=1}^{30}A_i=\{0,2,4,6,\ldots,60\}$, $\bigcap_{i=1}^{30}A_i=\{0\}$

(4) $\bigcup_{n=1}^{\infty}B_n=\{\ldots,-6,-4,-2,0,2,4,6,\ldots\}$, $\bigcap_{n=1}^{\infty}B_n=\{0\}$

(7) $\bigcup_{n=1}^{\infty}B_n=[-1,1)$, $\bigcap_{n=1}^{\infty}B_n=[-1,0]$

(10) $\bigcup_{t\in\mathbb{R}}C_t=[0,\infty)$, $\bigcap_{t\in\mathbb{R}}C_t=\emptyset$

(17) *Proof.* To show $B-(\bigcup_{i\in I}A_i)\subseteq\bigcap_{i\in I}(B-A_i)$, let $x\in B-(\bigcup_{i\in I}A_i)$. This means $x\in B$ and $x\notin(\bigcup_{i\in I}A_i)$. Thus we know that $x\notin A_i$ for all $i\in I$. Since $x\in B$ and $x\notin A_i$, for all $i\in I$, it follows that $x\in B-A_i$, for all $i\in I$. This yields $x\in\bigcap_{i\in I}(B-A_i)$.

To show $\bigcap_{i\in I}(B-A_i)\subseteq B-(\bigcup_{i\in I}A_i)$, let $y\in\bigcap_{i\in I}(B-A_i)$. Thus we have $y\in B-A_i$ for all $i\in I$. Thus for each $i\in I$, we know $y\in B$ and $y\notin A_i$. Given that $y\notin A_i$, for all $i\in I$, it follows that $y\notin\bigcup_{i\in I}A_i$. This means $y\in B-(\bigcup_{i\in I}A_i)$. □

(19) *Proof.* To show $(\bigcap_{i\in I}A_i)-B\subseteq\bigcap_{i\in I}(A_i-B)$, let $x\in(\bigcap_{i\in I}A_i)-B$. Thus we have $x\in\bigcap_{i\in I}A_i$ and $x\notin B$. Given that $x\in\bigcap_{i\in I}A_i$, it

follows that $x \in A_i$, for all $i \in I$. Thus for every $i \in I$, we know $x \in A_i$ and $x \notin B$. This implies $x \in A_i - B$ for all $i \in I$. It then follows that $x \in \bigcap_{i \in I} (A_i - B)$.

To show $\bigcap_{i \in I} (A_i - B) \subseteq (\bigcap_{i \in I} A_i) - B$, let $y \in \bigcap_{i \in I} (A_i - B)$. This means $y \in A_i - B$, for all $i \in I$. Thus for every $i \in I$, we know $y \in A_i$ and that $y \notin B$. This yields $y \in \bigcap_{i \in I} A_i$ and $y \notin B$, implying $y \in (\bigcap_{i \in I} A_i) - B$. □

Chapter 4 Exercises

Section 4.2

(1) *Proof.* Let $P(n)$ denote the statement $2 + 4 + 6 + \cdots + 2n = n(n+1)$ for all $n \in \mathbb{N}$.

Base Case: We must show $P(1)$ is true. When $n = 1$, the sum is 2 and $n(n+1) = 1(1+1) = 2$. This means $P(1)$ is true.

Inductive Case: Let $k \in \mathbb{N}$ and assume $P(k)$ is true. Thus we are assuming

$$2 + 4 + 6 + \cdots + 2k = k(k+1).$$

We must show $P(k+1)$ is true or that

$$2 + 4 + 6 + \cdots + 2k + 2(k+1) = (k+1)((k+1)+1) = (k+1)(k+2).$$

The calculation

$$
\begin{aligned}
2 + 4 + 6 + \cdots + 2k + 2(k+1) &= (2 + 4 + 6 + \cdots + 2k) + 2(k+1) \\
&= k(k+1) + 2(k+1) \\
&= (k+1)(k+2)
\end{aligned}
$$

shows the implication $P(k)$ implies $P(k+1)$ is true. □

(4) *Proof.* Let $P(n)$ denote the statement $4 + 10 + 16 + \cdots + 6n - 2 = n(3n+1)$ for all $n \in \mathbb{N}$.

Base Case: We must show $P(1)$ is true. When $n = 1$, the sum is 4 and $n(3n+1) = 1(3 \cdot 1 + 1) = 4$. Thus we have $P(1)$ is true.

Inductive Case: Let $k \in \mathbb{N}$ and assume $P(k)$ is true. Thus we are assuming

$$4 + 10 + 16 + \cdots + 6k - 2 = k(3k+1).$$

We must show $P(k+1)$ is true or that

$$4 + 10 + 16 + \cdots + 6(k+1) - 2 = (k+1)(3(k+1)+1) = (k+1)(3k+4).$$

The calculation

$$4 + 10 + 16 + \cdots + 6k - 2 + 6(k+1) - 2$$
$$= (4 + 10 + 16 + \cdots + 6k - 2) + 6(k+1) - 2$$
$$= k(3k+1) + 6(k+1) - 2 = 3k^2 + k + 6k + 6 - 2$$
$$= 3k^2 + 7k + 4 = (k+1)(3k+4)$$

shows the implication $P(k)$ implies $P(k+1)$ is true. □

(7) *Proof.* Let $P(n)$ denote the statement $2^2 + 4^2 + 6^2 + \cdots + (2n)^2 = \frac{2n(2n+1)(2n+2)}{6}$ for all $n \in \mathbb{N}$.

Base Case: We must show $P(1)$ is true. When $n = 1$, the sum is $2^2 = 4$ and $\frac{2n(2n+1)(2n+2)}{6} = \frac{2 \cdot 1(2 \cdot 1 + 1)(2 \cdot 1 + 2)}{6} = \frac{2 \cdot 3 \cdot 4}{6} = 4$. Since the two values are equal, $P(1)$ is true.

Inductive Case: Let $k \in \mathbb{N}$ and assume $P(k)$ is true. This means we are assuming

$$2^2 + 4^2 + 6^2 + \cdots + (2k)^2 = \frac{2k(2k+1)(2k+2)}{6}.$$

We must show $P(k+1)$ is true or that

$$2^2 + 4^2 + 6^2 + \cdots + (2k)^2 + (2(k+1))^2$$
$$= \frac{2(k+1)(2(k+1) + 1)(2(k+1) + 2)}{6}$$
$$= \frac{(2k+2)(2k+3)(2k+4)}{6}.$$

The calculation

$$2^2 + 4^2 + 6^2 + \cdots + (2k)^2 + (2(k+1))^2$$
$$= (2^2 + 4^2 + 6^2 + \cdots + (2k)^2) + (2(k+1))^2$$
$$= \frac{2k(2k+1)(2k+2)}{6} + (2k+2)^2$$
$$= \frac{2k(2k+1)(2k+2) + 6(2k+2)^2}{6}$$
$$= \frac{(2k+2)[2k(2k+1) + 6(2k+2)]}{6}$$
$$= \frac{(2k+2)[4k^2 + 14k + 12]}{6}$$

$$= \frac{(2k+2)(2k+3)(2k+4)}{6}$$

shows the implication $P(k)$ implies $P(k+1)$ is true. □

(10) *Proof.* Let $P(n)$ denote the statement $\frac{1}{1\cdot2} + \frac{1}{2\cdot3} + \frac{1}{3\cdot4} + \cdots + \frac{1}{n(n+1)} = \frac{n}{n+1}$ for all $n \in \mathbb{N}$.

Base Case: We must show $P(1)$ is true. When $n = 1$, the sum is $\frac{1}{1\cdot2} = \frac{1}{2}$ and $\frac{n}{n+1} = \frac{1}{1+1} = \frac{1}{2}$. Thus we have $P(1)$ is true.

Inductive Case: Let $k \in \mathbb{N}$ and assume $P(k)$ is true. This means we are assuming

$$\frac{1}{1\cdot2} + \frac{1}{2\cdot3} + \cdots + \frac{1}{k(k+1)} = \frac{k}{k+1}.$$

We must show $P(k+1)$ is true or that

$$\frac{1}{1\cdot2} + \frac{1}{2\cdot3} + \cdots + \frac{1}{(k+1)((k+1)+1)} = \frac{k+1}{(k+1)+1} = \frac{k+1}{k+2}.$$

The calculation

$$\frac{1}{1\cdot2} + \frac{1}{2\cdot3} + \cdots + \frac{1}{k(k+1)} + \frac{1}{(k+1)((k+1)+1)}$$
$$= \frac{k}{k+1} + \frac{1}{(k+1)(k+2)} = \frac{k(k+2)+1}{(k+1)(k+2)}$$
$$= \frac{k^2+2k+1}{(k+1)(k+2)} = \frac{(k+1)^2}{(k+1)(k+2)} = \frac{k+1}{k+2}$$

shows the implication $P(k)$ implies $P(k+1)$ is true. □

(13) *Proof.* Let $P(n)$ denote the statement $3|(n^3 - n)$ for all $n \in \mathbb{N}$.

Base Case: We must show $P(1)$ is true. When $n = 1$, it follows that $n^3 - n = 1^3 - 1 = 1 - 1 = 0 = 0\cdot3$. Since 0 is an integer, $3|(n^3 - n)$ and $P(1)$ is true.

Inductive Case: Let $k \in \mathbb{N}$ and assume $P(k)$ is true. This means $3|(k^3 - k)$, implying there exists an integer l such that $k^3 - k = 3l$. We must show $P(k+1)$ is true or that $3|((k+1)^3 - (k+1))$. Since

$$(k+1)^3 - (k+1) = k^3 + 3k^2 + 3k + 1 - k - 1 = k^3 - k + 3k^2 + 3k$$
$$= 3l + 3k^2 + 3k = 3(l + k^2 + k),$$

where $l + k^2 + k$ is an integer, it follows that $3|((k+1)^3 - (k+1))$. This shows the implication $P(k)$ implies $P(k+1)$ is true. □

(16) *Proof.* Let $P(n)$ denote the statement $2^n > n$ for all $n \in \mathbb{Z}$ with $n \geq 0$.

Base Case: We must show $P(0)$ is true. When $n = 0$, it follows that $2^n = 2^0 = 1$. Since $1 > 0$, the statement is true for $n = 0$.

Inductive Case: Let $k \in \mathbb{Z}$, with $k \geq 0$, and assume $P(k)$ is true. This means we are assuming $2^k > k$. We must show $P(k+1)$ is true or that $2^{k+1} > k + 1$.

First note that $2^{k+1} = 2^1 2^k = 2 \cdot 2^k = 2^k + 2^k$. Using the fact $2^k > k$, we get that $2^{k+1} = 2^k + 2^k > k + 2^k$. Given that $k \geq 0$, it follows that $2^k \geq 2^0$ or that $2^k > 1$. We now have

$$2^{k+1} > k + 2^k \geq k + 1,$$

implying

$$2^{k+1} > k + 1.$$

This shows the implication $P(k)$ implies $P(k+1)$ is true. □

(19) *Proof.* Let $P(n)$ denote the statement $B - (A_1 \cup A_2 \cup \cdots \cup A_n) = (B - A_1) \cap (B - A_2) \cap \cdots \cap (B - A_n)$ for all $n \in \mathbb{N}$.

Base Case: We must show $P(1)$ is true. When $n = 1$, the statement $B - A_1 = B - A_1$ is true by Lemma 3.2.4 (i).

Inductive Case: Let $k \in \mathbb{N}$ and assume $P(k)$ is true. Therefore, we are assuming

$$B - (A_1 \cup A_2 \cup \cdots \cup A_k) = (B - A_1) \cap (B - A_2) \cap \cdots \cap (B - A_k).$$

We must show $P(k+1)$ is true or that

$$B - (A_1 \cup A_2 \cup \cdots \cup A_k \cup A_{k+1}) = (B - A_1) \cap (B - A_2) \cap \cdots \cap (B - A_{k+1}).$$

First note that by rewriting, we have

$$B - (A_1 \cup A_2 \cup \cdots \cup A_{k+1}) = B - ((A_1 \cup A_2 \cup \cdots \cup A_k) \cup A_{k+1}).$$

Now apply Theorem 3.3.4 (vii) to the two sets $A_1 \cup A_2 \cup \cdots \cup A_k$ and A_{k+1} to obtain

$$B - ((A_1 \cup A_2 \cup \cdots \cup A_k) \cup A_{k+1})$$
$$= B - (A_1 \cup A_2 \cup \cdots \cup A_k) \cap (B - A_{k+1}).$$

By the induction hypothesis, we know

$$B - (A_1 \cup A_2 \cup \cdots \cup A_k) = (B - A_1) \cap (B - A_2) \cap \cdots \cap (B - A_k).$$

Using it, we get

$$B - (A_1 \cup A_2 \cup \cdots \cup A_k \cup A_{k+1})$$
$$= B - (A_1 \cup A_2 \cup \cdots \cup A_k) \cap (B - A_{k+1})$$
$$= (B - A_1) \cap (B - A_2) \cap \cdots \cap (B - A_k) \cap (B - A_{k+1}).$$

This shows the implication $P(k)$ implies $P(k+1)$ is true. ☐

(22) *Proof.* Let $P(n)$ denote the statement $1 + 5 + 5^2 + \cdots + 5^n = \frac{5^{n+1}-1}{4}$ for all integers n with $n \geq 0$.

Base Case: We must show $P(0)$ is true. When $n = 0$, the sum is 1 and $\frac{5^{n+1}-1}{4} = \frac{5^{0+1}-1}{4} = \frac{4}{4} = 1$. It follows that $P(0)$ is true.

Inductive Case: Let $k \in \mathbb{Z}$, with $k \geq 0$, and assume $P(k)$ is true. Therefore, we are assuming

$$1 + 5 + 5^2 + \cdots + 5^k = \frac{5^{k+1} - 1}{4}.$$

We must show $P(k+1)$ is true or that

$$1 + 5 + 5^2 + \cdots + 5^k + 5^{k+1} = \frac{5^{(k+1)+1} - 1}{4} = \frac{5^{k+2} - 1}{4}.$$

The calculation

$$1 + 5 + 5^2 + \cdots + 5^k + 5^{k+1} = \frac{5^{k+1} - 1}{4} + 5^{k+1} = \frac{5^{k+1} - 1 + 4 \cdot 5^{k+1}}{4}$$
$$= \frac{5 \cdot 5^{k+1} - 1}{4} = \frac{5^{k+2} - 1}{4}$$

shows the implication $P(k)$ implies $P(k+1)$ is true. ☐

(25) *Proof.* Let $P(n)$ denote the statement $4^n > 3^n + 2$ for all integers n with $n \geq 2$.

Base Case: We must show $P(2)$ is true. When $n = 2$, it follows that $4^n = 4^2 = 16$ and $3^n + 2 = 3^2 + 2 = 11$. Since $16 > 11$, we have $4^n > 3^n + 2$, for $n = 2$, and that the statement is true for $n = 2$.

Inductive Case: Let $k \in \mathbb{Z}$, with $k \geq 0$, and assume $P(k)$ is true. This means we are assuming $4^k > 3^k + 2$. We must show $P(k+1)$ is true or that $4^{k+1} > 3^{k+1} + 2$.

First note that $4^{k+1} = 4^1 4^k = (3+1)4^k = 3 \cdot 4^k + 4^k$. Using the fact that $4^k > 3^k + 2$, we get

$$4^{k+1} = 3 \cdot 4^k + 4^k > 3(3^k + 2) + 4^k = 3^{k+1} + 4^k + 6.$$

Given that $6 > 2$, it follows that $4^k + 6 > 2$ or that

$$4^{k+1} > 3^{k+1} + 4^k + 6 > 3^{k+1} + 2.$$

This means $4^{k+1} > 3^{k+1} + 2$, showing that the implication $P(k)$ implies $P(k+1)$ is true. □

Section 4.3

(3) *Proof.* Let $P(n)$ denote the statement $a_n = 2^n$ for $n \in \mathbb{N}$.

Base Case: We must show $P(1)$ and $P(2)$ are true. When $n = 1$, we have $a_1 = 2$ and $2^n = 2^1 = 2$. When $n = 2$, we have $a_2 = 4$ and $2^2 = 4$. Thus $P(n)$ is true for $n = 1$ and $n = 2$.

Inductive Case: Let k be an integer, such that $k \geq 2$, and assume $P(i)$ is true for all integers i, with $1 \leq i \leq k$. Thus, we are assuming $a_i = 2^i$, for $1 \leq i \leq k$. We must show $P(k+1)$ is true or that $a_{k+1} = 2^{k+1}$. By the definition of the sequence, we start by observing that

$$a_{k+1} = 5a_{(k+1)-1} - 6a_{(k+1)-2} = 5a_k - 6a_{k-1}.$$

Since $1 \leq k - 1, k \leq k$, we have $a_k = 2^k$ and $a_{k-1} = 2^{k-1}$. This implies

$$a_{k+1} = 5a_k - 6a_{k-1} = 5 \cdot 2^k - 6 \cdot 2^{k-1} = 5 \cdot 2^k - 3 \cdot 2 \cdot 2^{k-1}$$
$$= 5 \cdot 2^k - 3 \cdot 2^k = (5-3)2^k = 2^{k+1}.$$

This shows the implication $P(i)$, for $1 \leq i \leq k$, implies $P(k+1)$ is true. □

(6) *Proof.* Let $P(n)$ denote the statement $a_n = n^2$ for $n \in \mathbb{N}$.

Base Case: We must show $P(1), P(2)$, and $P(3)$ are true. When $n = 1$, we have $a_1 = 1$ and $1^2 = 1$. When $n = 2$, we have $a_2 = 4$ and $2^2 = 4$. And when $n = 3$, we have $a_3 = 9$ and that $3^2 = 9$. Thus $P(n)$ is true for $n = 1, n = 2$, and $n = 3$.

Inductive Case: Let k be an integer, such that $k \geq 3$, and assume $P(i)$ is true for all integers i, where $1 \leq i \leq k$. Thus we are assuming $a_i = i^2$ for $1 \leq i \leq k$. We must show $P(k+1)$ is true or that $a_{k+1} = (k+1)^2$.

By the definition of the sequence, we start by observing

$$a_{k+1} = a_{(k+1)-1} - a_{(k+1)-2} + a_{(k+1)-3} + 4(k+1) - 6$$
$$= a_k - a_{k-1} + a_{k-2} + 4k - 2.$$

Since $1 \leq k-2, k-1, k \leq k$, we have $a_k = k^2, a_{k-1} = (k-1)^2$, and $a_{k-2} = (k-2)^2$. This implies

$$a_{k+1} = a_k - a_{k-1} + a_{k-2} + 4k - 2$$
$$= k^2 - (k-1)^2 + (k-2)^2 + 4k - 2$$
$$= k^2 - k^2 + 2k - 1 + k^2 - 4k + 4 + 4k - 2$$
$$= k^2 + 2k + 1 = (k+1)^2.$$

This shows the implication $P(i)$, for $1 \leq i \leq k$, implies $P(k+1)$ is true. □

(7) HINT: Use the fact that $r^2 - r - 1 = 0$, which implies $r^2 = r + 1$. Similarly, it also follows that $s^2 = s + 1$.

(9) *Proof.* For $n \in \mathbb{N}$, with $n \geq 2$, let $P(n)$ denote the statement that n is prime or can be expressed as a product of primes.

Base Case: We must show $P(2)$ is true. This follows given that 2 is prime.

Inductive Case: Let k be an integer, such that $k \geq 2$, and assume $P(i)$ is true for all integers i, with $2 \leq i \leq k$. This means we are assuming i is prime or can be written as a product of primes, for $2 \leq i \leq k$. We must show $P(k+1)$ is true.

Now consider the natural number $k+1$. If $k+1$ is prime, then we are done. Proceed assuming that $k+1$ is not prime. Then $k+1$ is composite and $k+1 = ab$ with $a, b \in \mathbb{N}$ and $2 \leq a, b \leq k$. We know by the induction hypothesis that a and b are prime or can be written as a product of primes. This implies that $k+1 = ab$ can be expressed as a product of primes. This shows the implication $P(i)$, for $2 \leq i \leq k$, implies $P(k+1)$ is true. □

Chapter 5 Exercises

Section 5.1

(1) $R = \{(1,6),(1,8),(1,10),(2,6),(2,8),(2,10),(3,8),(3,10),(4,10)\}$
$R^{-1} = \{(6,1),(6,2),(8,1),(8,2),(8,3),(10,1),(10,2),(10,3),(10,4)\}$

(4) $R = \{(-2,4),(-1,1),(0,0),(1,1),(2,4)\}$

(7) (a) $(-4,2),(-3,1),(-2,-6),(-1,9),(0,2),(1,3),(2,-4),(3,3)$

  (b) An element $x \in \mathbb{Z}$ satisfies $x\,R\,0$ if and only if $x + 0$ is even. This implies x is even and that the set of all elements $x \in \mathbb{Z}$ that satisfy $x\,R\,0$ is the set of even integers $\{\ldots,-6,-4,-2,0,2,4,6,\ldots\}$.

  (c) An element $y \in \mathbb{Z}$ satisfies $y\,R\,1$ if and only if $y+1$ is even. This implies $y + 1 = 2n$ for $n \in \mathbb{N}$. This implies

$$y = 2n - 1 = 2n - 2 + 2 - 1 = 2(n-1) + 1,$$

  where $n - 1$ is an integer. Thus the set of all elements $y \in \mathbb{Z}$ that satisfy $y\,R\,1$ is the set of odd integers $\{\ldots,-5,-3,-1,1,3,5,\ldots\}$.

(10) $\{(3,4),(5,12),(8,15),(7,24),(20,21)\}$

Section 5.2

(1) (a) It is not reflexive, not symmetric, and not transitive.

  (d) It is symmetric, but it is not reflexive and not transitive.

  (g) It is symmetric and transitive, but it is not reflexive.

  (j) It is vacuously symmetric and transitive, but it is not reflexive.

(2) (b) This relation is not reflexive. For an integer a, it follows that $a + a = 2a$, which is always even. The relation is symmetric. If integers a and b satisfy $a\,R\,b$, meaning $a + b$ is odd, then given that $a+b = b+a$, we know that $b+a$ is odd and that $b\,R\,a$. This relation is not transitive. We know $1\,R\,2$ and $2\,R\,3$ given that $1+2 = 3$ and $2+3 = 5$ are both odd. But $1\slashed{R}3$ since $1 + 3 = 4$, which is not odd.

(d) This relation is reflexive. For an integer a, it follows that $a+4a = 5a$, implying that $5|(a+4a)$. Thus we have $a\,R\,a$. The relation is symmetric. Consider integers a and b satisfying $a\,R\,b$. This means $5|(a+4b)$, implying that $a+4b = 5l$, for some integer l, or that $a = 5l - 4b$. Then we have

$$b + 4a = b + 4(5l - 4b) = 20l - 15b = 5(4l - 3b),$$

where $4l - 3b$ is an integer. This means $5|(b+4a)$ and that $b\,R\,a$. This relation is transitive. Consider integers a, b, and c such that $a\,R\,b$ and $b\,R\,c$. Then we have $5|(a+4b)$ and $5|(b+4c)$, implying that $a + 4b = 5l$ and $b + 4c = 5k$, for integers l and k. Then we have

$$a + 4b + b + 4c = 5l + 5k$$
$$a + 4c = 5l + 5k - 5b$$
$$a + 4c = 5(l + k - b),$$

where $l + k - b$ is an integer. This means $5|(a + 4c)$ and $a\,R\,c$.

(f) This relation is not reflexive. For the natural number 2, it follows that $2+2 = 4$, which is not prime. The relation is symmetric. If integers a and b satisfy $a\,R\,b$, meaning $a+b$ is a prime number, then given that $a + b = b + a$, we know that $b + a$ is prime and that $b\,R\,a$. This relation is not transitive. We know $1\,R\,2$ and $2\,R\,3$ given that $1 + 2 = 3$ and $2 + 3 = 5$ are both prime. But $1\not\!R\,3$ since $1 + 3 = 4$, which is not prime.

(h) This relation is reflexive. For $x \in \mathbb{R}$, we have $xx = x^2 \geq 0$, implying $x\,R\,x$. The relation is symmetric. Consider $x, y \in \mathbb{R}$ satisfying $x\,R\,y$, meaning $xy \geq 0$. Since $xy = yx$, it follows that $yx \geq 0$. This implies $y\,R\,x$. This relation is not transitive. We know $-1\,R\,0$ and $0\,R\,1$ given that $(-1){\cdot}0 = 0 \geq 0$ and $0{\cdot}1 = 0 \geq 0$. But $-1\not\!R\,1$ since $(-1){\cdot}1 = -1 < 0$.

Section 5.3

(1) (a)   i. The set $\mathcal{P}_1$ is a partition of set A.

    ii. The set $\mathcal{P}_2$ is not a partition of set A.

    iii. The set $\mathcal{P}_3$ is not a partition of set A.

(3) (a) This is not a partition of $\mathbb{R}$ as -1 is in both subsets of $\mathbb{R}$ in $\mathcal{P}_1$.

(b) This is a partition of $\mathbb{R}$ (see Definition 2.3.10).

(5) (c) This is not an equivalence relation. It is reflexive and transitive, but is not symmetric. Given the integers 3 and 1, we know $3\,R\,1$ since $3 - 1 = 2 \geq 0$. However, $1\,\cancel{R}\,3$ given that $1 - 3 = -2 \not\geq 0$.

(f) This relation is an equivalence relation. For any $x \in \mathbb{R}$, we have $x - x = 0$ with $0 \in \mathbb{Q}$. This means $x\,R\,x$ and that R is reflexive. Now let $x, y \in \mathbb{R}$ and assume that $x\,R\,y$. This means $x - y \in \mathbb{Q}$, yielding $x - y = \frac{a}{b}$, for $a, b \in \mathbb{Z}$ with $b \neq 0$. Then we have $(-1)(x - y) = (-1)\frac{a}{b}$, implying $y - x = \frac{-a}{b}$. Thus $y - x \in \mathbb{Q}$, resulting in $y\,R\,x$. Thus R is symmetric. To show R is transitive, let $x, y, z \in \mathbb{R}$ such that $x\,R\,y$ and $y\,R\,z$. This means $x - y, y - z \in \mathbb{Q}$. Given that $(x - y) + (y - z) = x - z$ and that the sum of two rational numbers is rational (you can prove this directly or use Theorem 2.4.7), we have $x - z \in \mathbb{Q}$. This results in $x\,R\,z$.

(6) (d) Let $a \in \mathbb{Z}$. Given that $a + 2a = 3a$, with a an integer, $3|(a + 2a)$ and $a\,R\,a$. Thus R is reflexive. Now let $a, b \in \mathbb{Z}$ and assume that $a\,R\,b$. This means $3|(a + 2b)$ or that $a + 2b = 3k$ for some integer k. Since $a = 3k - 2b$, we have

$$b + 2a = b + 2(3k - 2b) = 6k - 3b = 3(2k - b),$$

where $2k - b$ is an integer. This implies $3|(b + 2a)$ and that $b\,R\,a$. This means R is symmetric. Next, let $a, b, c, \in \mathbb{Z}$ and assume $a\,R\,b$ and $b\,R\,c$. This implies $a + 2b = 3k$ and $b + 2c = 3l$ for integers k and l. Then we have

$$a + 2b + b + 2c = 3k + 3l$$
$$a + 2c = 3k + 3l - 3b = 3(k + l - b)$$

where $k + l - b$ is an integer. This means $3|(a + 2c)$ and that $a\,R\,c$. Thus R is transitive.

$$[0] = \{x \in \mathbb{Z} : 3|x\} = \{\ldots, -9, -6, -3, 0, 3, 6, 9, \ldots\}$$
$$[1] = \{x \in \mathbb{Z} : 3|(x + 2)\} = \{\ldots, -8, -5, -2, 1, 4, 7, 10, \ldots\}$$
$$[2] = \{x \in \mathbb{Z} : 3|(x + 4)\} = \{\ldots, -7, -4, -1, 2, 5, 8, 11, \ldots\}$$

(9) $[\frac{1}{5}] = \{x \in \mathbb{Q} : x - \frac{1}{5} \in \mathbb{Z}\} = \{\ldots, \frac{-14}{5}, \frac{-9}{5}, \frac{-4}{5}, \frac{1}{5}, \frac{6}{5}, \frac{11}{5}, \ldots\}$

Chapter 6 Exercises

Section 6.2

(1) (a)   i. The relation R_1 is a function.

   ii. The relation R_2 is not a function.

   iii. The relation R_3 is not a function.

   (c) This relation is a function. (Rewrite $2n - m = 1$.)

   (e) This relation is a function. For $n \in \mathbb{Z}$, since $n^2 + 1 \in \mathbb{Z}$, with $n^2 + 1 \geq 1$, it follows that $m = n^2 + 1 \in \mathbb{N}$.

   (g) This relation is a function. For every rational number $\frac{a}{b}$, with $a, b \in \mathbb{Z}$, with $b \neq 0$, the product ab results in only one integer.

   (i) This relation is not a function. The element $0 \in \mathbb{R}$ is related to both of the real numbers -1 and 1.

(5) (a)   i. $f(\{0, 4, 8\}) = \{b, d\}$.   iii. $f^{-1}(\{b, e\}) = \{0, 8\}$.

   ii. $\operatorname{ran}(f) = \{b, c, d\}$.

   (c)   i. $f(\{-2, -1, 0, 1, 2\}) = \{-4, -1, 2, 5, 8\}$.

   ii. $\operatorname{ran}(f) = \{\ldots, -10, -7, -4, -1, 2, 5, 8, 11, 14, \ldots\}$
   $= \{m \in \mathbb{Z} : m \pmod 3 = 2\}$

   (e)   i. $f(\{7m : m \in \mathbb{Z}\}) = \{0\}$.

   ii. $\operatorname{ran}(f) = \{0, 1, 2, 3, 4, 5, 6\}$.

   iii. $f^{-1}(\{3\}) = \{\ldots, -25, -18, -11, -4, 3, 10, 17, 24, \ldots\}$
   $= \{m \in \mathbb{Z} : m \pmod 7 = 3\}$.

   iv. $f^{-1}(\{5\}) = \{\ldots, -23, -16, -9, -2, 5, 12, 19, 26, \ldots\}$
   $= \{m \in \mathbb{Z} : m \pmod 7 = 5\}$

   (g)   i. $f([-2, 2]) = [1, 5]$.

   ii. $\operatorname{ran}(f) = [1, \infty)$.

   iii. $f^{-1}([9, \infty)) = (-\infty, -\sqrt{8}] \cup [\sqrt{8}, \infty)$.

(7) *Proof.* Assume that $f(\emptyset) \neq \emptyset$. Since $f(\emptyset) \subseteq B$, there exists an element $b \in B$ such that $b \in f(\emptyset)$. This implies that there exists an element $a \in \emptyset$ such that $f(a) = b$. This is a contradiction as the empty set contains no elements. This results in $f(\emptyset) = \emptyset$.   □

(12) *Proof.* Let $P(n)$ denote the statement $f^{-1}(Y_1 \cap Y_2 \cap \cdots \cap Y_n) = f^{-1}(Y_1) \cap f^{-1}(Y_2) \cap \cdots \cap f^{-1}(Y_n)$ for all $n \in \mathbb{N}$.

Base Case: We must show that $P(1)$ is true. When $n = 1$, the statement $f^{-1}(Y_1) = f^{-1}(Y_1)$ is true by Lemma 3.2.4 (i).

Inductive Case: Let $k \in \mathbb{N}$. Assume $P(k)$ is true or that

$$f^{-1}(Y_1 \cap Y_2 \cap \cdots \cap Y_k) = f^{-1}(Y_1) \cap f^{-1}(Y_2) \cap \cdots \cap f^{-1}(Y_k).$$

We must show that $P(k+1)$ is true or that

$$f^{-1}(Y_1 \cap Y_2 \cap \cdots \cap Y_{k+1}) = f^{-1}(Y_1) \cap f^{-1}(Y_2) \cap \cdots \cap f^{-1}(Y_{k+1}).$$

First note that

$$f^{-1}(Y_1 \cap Y_2 \cap \cdots \cap Y_{k+1}) = f^{-1}(Y_1 \cap Y_2 \cap \cdots \cap Y_k \cap Y_{k+1})$$
$$= f^{-1}((Y_1 \cap Y_2 \cap \cdots \cap Y_k) \cap Y_{k+1}).$$

Use the induction hypothesis and Proposition 6.2.12 (iii), applied to the two sets $Y_1 \cap Y_2 \cap \cdots \cap Y_k$ and Y_{k+1}, to obtain

$$f^{-1}(Y_1 \cap Y_2 \cap \cdots \cap Y_{k+1})$$
$$= f^{-1}((Y_1 \cap Y_2 \cap \cdots \cap Y_k) \cap Y_{k+1})$$
$$= f^{-1}(Y_1 \cap Y_2 \cap \cdots \cap Y_k) \cap f^{-1}(Y_{k+1})$$
$$= f^{-1}(Y_1) \cap f^{-1}(Y_2) \cap \cdots \cap f^{-1}(Y_k) \cap f^{-1}(Y_{k+1}).$$

This shows that the implication $P(k)$ implies $P(k+1)$ is true. □

(17) *Proof.* To show $f^{-1}(Y_1 - Y_2) \subseteq f^{-1}(Y_1) - f^{-1}(Y_2)$, let $a \in f^{-1}(Y_1 - Y_2)$. This means that $f(a) \in Y_1 - Y_2$, implying that $f(a) \in Y_1$ and $f(a) \notin Y_2$. Given $f(a) \in Y_1$, it follows that $a \in f^{-1}(Y_1)$. If $a \in f^{-1}(Y_2)$, then we would have $f(a) \in Y_2$, a contradiction. Consequently, we know $a \in f^{-1}(Y_1)$ and $a \notin f^{-1}(Y_2)$, yielding $a \in f^{-1}(Y_1) - f^{-1}(Y_2)$.

To show and $f^{-1}(Y_1) - f^{-1}(Y_2) \subseteq f^{-1}(Y_1 - Y_2)$, let $a \in f^{-1}(Y_1) - f^{-1}(Y_2)$. It follows that $a \in f^{-1}(Y_1)$ and $a \notin f^{-1}(Y_2)$. This means $f(a) \in Y_1$ and that $f(a) \notin Y_2$. As a result, we have $f(a) \in Y_1 - Y_2$ and that $a \in f^{-1}(Y_1 - Y_2)$. □

Section 6.3

(1) (a) i. The function f_1 is not one-to-one and not onto.

ii. The function f_2 is not one-to-one and not onto.

(d) This function is not one-to-one as $f(-1) = f(1)$. This function is onto. Let $m \in \mathbb{N}$ and consider the integer $m - 1$. First note that since $m \geq 1$, it follows that $m - 1 \geq 0$. Then we have $f(m - 1) = |m - 1| + 1 = m - 1 + 1 = m$.

(f) This function is one-to-one. Let $n, m \in \mathbb{Z}$ and assume that $f(n) = f(m)$. This implies $(n - 1, n + 1) = (m - 1, m + 1)$. For the ordered pairs to be equal, we must have $n - 1 = m - 1$ (and $n + 1 = m + 1$). This implies $n = m$. This function is not onto. Consider the element $(-2, 1) \in \mathbb{Z} \times \mathbb{Z}$. To be onto, there must exist an $n \in \mathbb{Z}$ such that $f(n) = (-2, 1)$. This means $(n - 1, n + 1) = (-2, 1)$, requiring $n - 1 = -2$ and $n + 1 = 1$. This results in $n = -1$ and $n = 0$, which is a contradiction.

(h) This function is not one-to-one as $f(-1) = f(1)$. This function is not onto. For $x \in \mathbb{R}$, we have $x^2 \geq 0$ and $x^2 + 1 \geq 1$. This means $f(x) = \sqrt{x^2 + 1} \geq 1$. Thus there is no $x \in \mathbb{R}$ such that $f(x) = -3$.

(3) *Proof.* ($\Rightarrow$) Let b be an element in B. Since f is onto, there exists an element $a \in A$ such that $f(a) = b$. Thus we have $a \in f^{-1}(\{b\})$. To show that a is unique, suppose there exists another element $a_1 \in A$ such that $a_1 \in f^{-1}(\{b\})$. This means $f(a_1) = b = f(a)$ or that $f(a_1) = f(a)$. Given that f is one-to-one, it follows that $a = a_1$ and that a is unique. This results in $f^{-1}(\{b\}) = \{a\}$.

($\Leftarrow$) To show f is one-to-one, let $a_1, a_2 \in A$ and assume that $f(a_1) = f(a_2)$. Since $f(a_1) \in B$, there exists an element $b \in B$ such that $f(a_1) = b$. The fact that $f(a_1) = f(a_2) = b$ implies that $a_1, a_2 \in f^{-1}(\{b\})$. This results in $a_1 = a_2$ and that f is one-to-one.

To show f is onto, let $b \in B$. By our hypothesis, $f^{-1}(\{b\}) = \{a\}$ for $a \in A$. This means $f(a) = b$ and that f is onto. $\square$

(5) *Proof.* To show $f(f^{-1}(Y)) \subseteq Y$, let $y \in f(f^{-1}(Y))$. Thus there is an $x \in f^{-1}(Y)$ such that $f(x) = y$. Given that $x \in f^{-1}(Y)$, it follows that $f(x) \in Y$ and that $y \in Y$.

To show $Y \subseteq f(f^{-1}(Y))$, let $y \in Y$. Since f is onto, there exists an element $x \in A$ such that $f(x) = y$. This means $f(x) \in Y$ and that $x \in f^{-1}(Y)$. It then follows that $y = f(x) \in f(f^{-1}(Y))$. $\square$

Section 6.4

(1) (a) $g \circ f = \{(0, -4), (2, -1), (4, -1), (6, -3), (8, -1)\}$

(d) For $(n, m) \in \mathbb{Z} \times \mathbb{Z}$, we have

$$(g \circ f)((n, m)) = g(f((n, m))) = g(n + m) = (-(n + m), n + m).$$

For $t \in \mathbb{Z}$, we have

$$(f \circ g)(t) = f(g(t)) = g((-t, t)) = -t + t = 0.$$

(g) $(h \circ g \circ f)(x) = h(g(f(x))) = h(g(2x)) = h(4x^2) = \sqrt{16x^4 + 1}$

$(f \circ f \circ f)(x) = f(f(f(x))) = f(f(2x)) = f(4x) = 8x$

$(f \circ g \circ h)(x) = f(g(h(x))) = f(g(\sqrt{x^2 + 1})) = f(x^2 + 1)$
$$= 2x^2 + 2$$

$(h \circ h \circ g)(x) = h(h(g(x))) = h(h(x^2)) = h(\sqrt{x^4 + 1})$
$$= \sqrt{x^4 + 2}$$

(4) (a) HINT: Let $a_1, a_2 \in A$ and assume that $f(a_1) = f(a_2)$. Now plug both values into g and use the fact that $g \circ f$ is one-to-one.

(c) HINT: Let $c \in C$ and begin by using the fact that $g \circ f : A \to C$ is onto.

Section 6.5

(1) (a) i. This function is invertible.

ii. This function is not invertible, as it is not one-to-one.

(d) This function is not invertible, as it is not one-to-one ($f(1) = f(-1)$.)

(g) This function is not invertible. While it is one-to-one, it is not onto. Let $(a, b) \in \mathbb{Z} \times \mathbb{Z}$, and suppose there exists an $(n, m) \in \mathbb{Z} \times \mathbb{Z}$ such that $f((n, m)) = (a, b)$. Then we have $(n - m, n + m) = (a, b)$ or that $n - m = a$ and $n + m = b$. Adding, we get $2n = a + b$, implying that $a + b$ is even. Thus for the element $(1, 2) \in \mathbb{Z} \times \mathbb{Z}$, given that $1 + 2 = 3$, where 3 is odd, there is no element $(n, m) \in \mathbb{Z} \times \mathbb{Z}$ such that $f((n, m)) = (1, 2)$.

(2) (a) Given that

$$(g \circ f)(n) = g(f(n)) = g(n - 2) = 2 + n - 2 = n = i_{\mathbb{Z}}(n)$$

and

$$(f \circ g)(n) = f(g(n)) = g(2 + n) = 2 + n - 2 = n = i_{\mathbb{Z}}(n),$$

the functions f and g are inverses to each other.

(c) Given that

$$(g \circ f)(x) = g(f(x)) = g(x^3 + 1) = (x^3 + 1 - 1)^3 = x^6 \neq i_\mathbb{R}(x),$$

we have that f and g are not inverses to each other.

(4) (a) HINT: let $b \in B$ and show that $(f^{-1} \circ f \circ f^{-1})(b) = f^{-1}(b)$, recalling that $f^{-1}(b) = a$ where $a \in A$ and $f(a) = b$.

Chapter 7 Exercises
Section 7.2

(1) (a) HINT: Show that the function $f : \{1, \ldots, n\} \to \{-1, \ldots, -n\}$ defined by $f(a) = -a$, for $a \in \{1, \ldots, n\}$, is a bijection.

  (d) HINT: Show that the function $f : \mathbb{Z} \to O$ defined by $f(n) = 2n + 1$, for $n \in \mathbb{Z}$, is a bijection.

  (g) HINT: Find an exponential function that will work.

  (h) HINT: Consider the function $f : (0, 1) \to (a, b)$ defined by $f(x) = (b - a)x + a$ for all $x \in (0, 1)$.

(2) HINT: Use Exercises (4) and (5) from Section 6.3.

(3) HINT: Use Exercise 1(h) and Theorem 7.2.3.

Section 7.3

(1) (a) $|A \times B \times A| = |A \times B| \cdot |A| = |A| \cdot |B| \cdot |A| = n^2 m$

  (d) $|\mathcal{P}(A \times B)| = 2^{|A \times B|} = 2^{|A| \cdot |B|} = 2^{nm}$

  (g) $|\mathcal{P}(\emptyset) \times \mathcal{P}(\mathcal{P}(\emptyset))| = |\mathcal{P}(\emptyset)| \cdot |\mathcal{P}(\mathcal{P}(\emptyset))| = 2^0 \cdot 2^{|\mathcal{P}(\emptyset)|} = 1 \cdot 2^1 = 2$

(4) HINT: Let $P(n)$, for $n \geq 0$, be the statement that if B is a finite set, with $|B| = n$, and $A \subseteq B$, then A is finite with $|A| \leq |B|$. For the inductive case, with $|B| = k + 1$, for $k \geq 0$, let $b \in B$ and consider the subset $B - \{b\}$. You will need Lemma 7.3.2.

(5) (a) HINT: Use Lemma 6.3.4 and Theorem 7.3.3.

(6) *Proof.* Recall by Exercise 19 in Section 3.3 that $A \cup B = A \cup (B - A)$ with $A \cap (B - A) = \emptyset$. Given that $B - A \subseteq B$, with B finite, we know by Theorem 7.3.3 that $B - A$ is finite. Let $|A| = n$ and $|B - A| = m$. Then there exist bijective functions $f : A \to \{1, 2, \ldots, n\}$ and $g : B - A \to$

$\{1, 2, \ldots, m\}$. Define a function $h : A \cup B \to \{1, 2, \ldots, n + m\}$ as follows:

$$h(x) = \begin{cases} f(x) & \text{if } x \in A \\ n + g(x) & \text{if } x \in B - A \end{cases}$$

Showing that h is a bijection is similar to the work done in the proof of Lemma 7.3.2. This implies that $|A \cup B| = n + m$ and that $A \cup B$ is finite. □

Section 7.4

(1) HINT: To show f is one-to-one, let $n, m \in \mathbb{N}$ and assume $f(n) = f(m)$. Examine the two cases of n and m having the same and opposite parity. To show f is onto, let $t \in \mathbb{Z}$ and examine the three cases of $t < 0, t = 0$, and $t > 0$.

(3) HINT: First note that $A \cup B = A \cup (B - A)$ with $A \cap (B - A) = \emptyset$. Since A is countably infinite, there exists a bijection $f : \mathbb{N} \to A$. We know by Theorem 7.4.3 that $B - A$ is countable. If $B - A$ is finite, there is a bijection $h_1 : \{1, 2, \ldots, n\} \to B - A$, for $n \in \mathbb{N}$. If $B - A$ is infinite, there is a bijection $h_2 : \mathbb{N} \to B - A$. There are two cases to examine. In each case, use f and h_1 or f and h_2 to construct a bijection from $\mathbb{N}$ to $A \cup B$.

(5) Use Theorem 7.4.6 and induction.

Section 7.5

(1) HINT: Assume $A \times B$ or $A \cup B$ is countable and reach a contradiction.

(3) HINT: Let $a \in A$. First show that $\{a\} \times \mathbb{R}$ is uncountable. Then apply Lemma 7.5.2.

Section 7.6

(1) HINT: Use the fact that $\mathbb{Q}$ is countably infinite.

(3) *Proof.* Define a function $f : A \to A \cup B$ by $f(a) = a$. It follows directly that f is one-to-one. If there exists a bijection from A to $A \cup B$, then $|A| = |A \cup B|$. If no bijection exists, then function f implies $A < |A \cup B|$. □

Bibliography

[Dra89] S. Drake. *A History of Free Fall: Aristotle to Galileo, with an Epilogue on Pi in the Sky*. Walt & Thompson, Toronto, 1989.

[End77] Herbert B. Enderton. *Elements of Set Theory*. Academic Press, New York, 1977.

[FG87] J. Fauvel and J. Gray, editors. *The History of Mathematics: A Reader*. Palgrave Macmillion with The Open University, New York, 1987.

[Gal] Galileo Galilei. *Dialogues Concerning Two New Sciences*. Translated from the Italian and Latin into English by Henry Crew and Alfonso de Salvio with an introduction by Antonio Favaro, Macmillan, New York, 1914, 3/17/2020, <https://oll.libertyfund.org/titles/753>.

[GG98] I. Gratan-Guinness. *The Norton History of the Mathematical Sciences: The Rainbow of Mathematics*. W.W. Norton & Company, New York, 1998.

[GU89] William Gustason and Dolph E. Ulrich. *Elementary Symbolic Logic*. Waveland Press, Inc., Prospect Heights, Illinois, 2 edition, 1989.

[Hal17] Paul R. Halmos. *Naive Set Theory*. Dover Publications, Inc., Mineola, New York, 2017.

[Hea02] T. L. Heath, editor. *Euclid's Elements: All Thirteen Books in One Volume*. Green Lion Press, Santa Fe, 2002.

[JJ98] Gareth A. Jones and J. Mary Jones. *Elementary Number Theory*. Springer-Verlag, Berlin, 1998.

[O'L16] Michael L. O'Leary. *A First Course in Mathematical Logic and Set Theory*. John Wiley & Sons, Inc., Hoboken, 2016.

[PL07]   A. S. Posamentier and I. Lehmann. *The Fabulous Fibonacci Numbers*. Prometheus Books, Amherst, New York, 2007.

[SB92]   D. Schmandt-Besserat. *Before Writing, Volume I, From Coutning to Cuneiform*. University of Texas Press, Austin, 1992.

[Sup72] P. Suppes. *Axiomatic Set Theory*. Dover Publications, Inc., New York, 1972.

Index

Printed in the United States
By Bookmasters